PRAIS
THE (BIG) YEAR THAT FLEW BY

"Dutch birders take their pursuit to a higher plane of skill and intensity, as exemplified in this global trek by Arjan Dwarshuis. A fast-paced page-turner and a unique adventure story, *The (Big) Year that Flew By* is also filled with insights about landscapes, people, and a world of wonderful birds."

—KENN KAUFMAN, author of *Kingbird Highway*

"Arjan artistically weaves together the beauty of the birds, the importance of conservation, and the complex logistics of nonstop travel. The roller coaster highs of seeing a target bird, paired with the lows of absolute fatigue and exhaustion, are an innate undertone of any birder's big year."

—TIFFANY KERSTEN, birding guide and continental U.S. Big Year record holder

"*The (Big) Year that Flew By* is not simply a celebration of a broken record but a global call to action to protect the habitats that birds rely on for survival. A Big Year of birding is a massive undertaking, requiring a herculean effort to successfully plan and execute. While this quest involves daunting logistics and emotional hurdles, the most notable achievement is distilling so much lived experience into a single book. Arjan has conquered both, beautifully capturing remote wild places and conveying the intensity endured to find and observe two-thirds of the world's bird species. Arjan's contagious passion for birds infuses his account of a global avian scavenger hunt like no other."

—CHRISTIAN HAGENLOCHER, author of *Falcon Freeway*

"Arjan's story is brilliantly told. I was with him every step of the way. It is much more than just a story about one man's bid to see as most of the world's bird species in one year as humanly possible. No, this is an epic journey by a man who's not only obsessed with birds but who has a deep spiritual connection with the planet as he observes the environments and habitats he encounters. It is clear that we have to do more to take care of our world and all its inhabitants, including us."

—DAVID LINDO, author of *The Urban Birder*

"An astonishing achievement! Arjan Dwarshuis's year was so big it covered an entire planet. His eagle eyes took in almost 7,000 species of bird and brought into sharp focus their often-fragile existence. Dwarshuis's heartfelt prose reminds us that many of these species are living on the edge—just a generation or two away from extinction. Yet this is a story of hope—Dwarshuis shines a light on the many communities around the world banding together to save their local birdlife. This book is a glorious tribute to the wealth of beauty and diversity found in birds, and a clarion call for us all to care about the future of birds everywhere."

—NEIL HAYWARD, author of *Lost Among the Birds*

"*The (Big) Year that Flew By* is somehow both a fast-paced race through and an in-depth immersion in an amazing birding Big Year. Arjan Dwarshuis does an excellent job of portraying the excitement, stress, and exertion that are part of any Big Year, but that was even more so in his record-breaking worldwide Big Year. I very much enjoyed reading his account, sometimes with my heart pounding, as I lived his journey with him, feeling joy when he managed to find another rarity, concern with his bird-seeking struggles, and sorrow in his very rare misses. As I read, I continued to be impressed by the logistics required for all his travels and by the excellent people who helped him find all these bird species. His interspersing of tales of his early birding years helped me better understand the how and why he was able to pull off his remarkable feat as well as his passion for birding. Throughout this book, he also explained conservation concerns and advances as they relate to the birds he sought, providing words of wisdom and words of hope. In his conclusion, he admits that while flying all over the world does use much energy, ecotourism in many far-flung places has been and continues to be of great importance in saving birds and habitats. I especially appreciate his conclusion in which he asks his readers to adopt a positive attitude and do their part—'enjoy nature and look up to the sky.'"

—LYNN E. BARBER, author of *Extreme Birder*

THE (BIG) YEAR
THAT FLEW BY

THE (BIG) YEAR THAT FLEW BY

Twelve Months
Six Continents
and the
Ultimate Birding Record

Arjan Dwarshuis

TRANSLATED BY Els Vanbrabant
FOREWORD BY Mark Obmascik

CHELSEA GREEN PUBLISHING
White River Junction, Vermont
London, UK

Originally published in Dutch by Meulenhoff Boekerij bv in the Netherlands
as *Een bevlogen jaar*. Published by special arrangement with Meulenhoff Boekerij bv
in conjunction with their duly appointed agent, 2 Seas Literary Agency.

This edition published by Chelsea Green Publishing, 2023.

Project Manager: Rebecca Springer
Developmental Editor: Matthew Derr
Copy Editor: Diane Durrett
Proofreader: Angela Boyle
Designer: Melissa Jacobson
Page Layout: Abrah Griggs

Printed in the United States of America.
First printing February 2023.
10 9 8 7 6 5 4 3 2 1   23 24 25 26 27

ISBN 978-1-64502-191-9 (paperback) | ISBN 978-1-64502-192-6 (ebook)
| ISBN 978-1-64502-193-3 (audiobook)

Library of Congress Cataloging-in-Publication Data is available.

Chelsea Green Publishing
P.O. Box 4529
White River Junction, Vermont USA

Somerset House
London, UK

www.chelseagreen.com

For my mother.
For all your tender love and care and
your unconditional support.

It seems to me that the natural world is the greatest source of excitement; the greatest source of visual beauty; the greatest source of intellectual interest. It is the greatest source of so much in life that makes life worth living.

—Sir David Attenborough

CONTENTS

FOREWORD

If you had one year of your life to do whatever you wanted—no work or family obligations—what would you do? Some people might hang out on a beach, or golf, or visit wineries or baseball stadiums or art museums.

Then there's Arjan Dwarshuis.

He chased birds.

Not just in his neighborhood park but around the world—across six continents, forty-one countries, fifteen seas, and five oceans. With binoculars and spotting scopes, he followed birds through jungles and deserts, on beaches and atop mountains, in landfills and dark alleys and sewage lagoons. He welcomed the dawn chorus of songbirds nesting in prairies and wished a good night to owls and nightjars hunting in dark forests. He logged 140,000 kilometers by plane, train, automobile, horse, bus, canoe, bike, ferry, motorcycle, truck roof, and hiking boots. He rode two rickshaws, too, as well as a van that kept blowing out its radiator in the remote savannah of Uganda.

He braved cyclones in Australia, blizzards in Minnesota and Patagonia, monsoon in Vietnam, and mudslides in Guatemala. It's hard to say what gave him more of a fright—the tiger that killed two cows just outside his vehicle in India, the two burly toughs who sand-wiched him in the backseat of his transport in Papua New Guinea, or the malaria in Suriname!

Or maybe the scariest thing was really that iffy chicken tikka masala from the street vendor just south of the Himalayas. That one really hurt. For a while.

For one entire year, Arjan Dwarshuis was consumed by birds, and for one year, the birds nearly consumed him. He gave up his job as a bartender in Amsterdam—he had poured Heinkens in a joint across the street from the Anne Frank Museum—and was at home in his apartment for only 5 of 366 days that year. (He planned all this for a Leap Year to give himself an extra day of birding in the field.)

His beat-the-clock obsession was magnificent, grueling, and enlightening. Borderline crazy, too.

The result was something no human in history ever had done—Arjan Dwarshuis saw 6,852 avian species in one year, nearly two-thirds of all birds known to exist on earth.

It all amounted to something birders call a Big Year, a once-in-a-lifetime chance to release the brakes on your life and let yourself be engulfed in an unparalleled global adventure. Birders around the world dream of doing a Big Year. Dwarshuis did the biggest one ever—at age twenty-nine.

The logistics alone would be enough to drive most people mad. Do the math. To break the existing Big Year world record, Dwarshuis needed to see seventeen different species of birds every day. Seventeen per day, every day, all year long. That assumes zero days of rest.

What complicates matters is that many birds are extremely particular about where they roost. If you live in a city, you might be able to find a pigeon or starling just about anywhere. But the only place in the world to see a Horned Guan, for example, is a 100-mile stretch of undisturbed evergreen forest along the mountainous west coast of Guatemala and Mexico. The gigantic Philippine Eagle—also known as the Monkey Eagle, for its dietary preference—can be found, believe it or not, only in the Philippines. The Geomalia, a shy rufous ground hopper with no known song, has never strayed from the difficult Sulawesi montane rain forests of Indonesia.

None of these areas are easy to reach. And even if you can arrange travel to these remote avian habitats, there is no guarantee the birds will be waiting. Lynyrd Skynyrd was right—the world is populated with free birds, who come and go as they please. For every Gunnison Sage-Grouse, which typically remains within a few miles of its original nest, there is a Bar-tailed Godwit, which flies 8,000 miles nonstop over the Pacific Ocean during semiannual migratory trips from Alaska to Tasmania and back.

To see as many species in as little time as possible—remember his relentless 17 bird-per-day mandate—Dwarshuis placed high priority on visiting the geographic pinch points of Panama and Israel, where millions of birds funnel through each spring and fall. Bird migration remains one of the world's greatest natural phenomena, an incredible spectacle that, for the unaware, remains hidden in plain sight. Dwarshuis didn't just get to see it. He lived it.

Of course, none of this was cheap. But it wasn't as expensive as you might think, either. By shrewdly lining up corporate sponsorships, bird guide partnerships, international friendships—and a few well-timed loans from the Bank of Dad—Dwarshuis was able to complete his worldwide expedition for a total cost of $65,000.

As you sit in your comfortable chair at home, turning the pages of this book, think about that price. Maybe you, too, could scratch up that kind of money. Maybe you, too, could negotiate a year of global travel on the fly. Maybe you, too, could see some of the rarest and most gorgeous creatures in some of the most spectacular landscapes on earth.

Or maybe you could just enjoy reading about someone who had the heart to do it himself. You don't have to own binoculars to be inspired by this wild adventure story. You just need imagination and dreams.

—Mark Obmascik,
author of *The Big Year: A Tale of Man, Nature, and Fowl Obsession*

Los Tarrales Nature Reserve, Suchitepéquez, Guatemala (2016)

When the alarm went off, it took me a few seconds before I realized where I was. Then my thoughts fell into place, and my drowsiness disappeared in one fell swoop. It was two o'clock in the morning. *Game time.*

My father had come to see me. Wanting to join me at all costs, he had trained for this challenge for months on end. Before his departure, I had called my mother from Colombia. Her voice was full of concern when she told me, "Ar, your father has gone mad. He's been walking up and down the stairs all day. To be fit for your hunt to find some kind of turkey."

Just before he left, disaster struck when he raced up those stairs one more time. He stumbled and tore his calf muscle.

My father is definitely not an emotive guy, so when I saw the tears in his eyes at the Guatemala City airport, I knew there must be more to it than just a torn muscle.

In the middle of the night, while I get into the Toyota Land Cruiser with my guide, John, my father waves me off. When I turn around, I can just see his silver-gray hair being swallowed up by the darkness.

We drive at a walking pace; the road is almost as narrow as the car and riddled with deep mud puddles. The many bumps and potholes cause me to shake back and forth in the back seat. A big branch stretches out across the road, but the driver seems oblivious, and the wood gives way with a loud squeak when it cracks up under the tires. After more than an hour, we finally arrive at a small clearing in the middle of the jungle. The start of the *Sendero de Lágrimas*, the Path of Tears.

A husky Guatemalan with a machete and headlamp leads us up the steep path that winds its way through the dense vegetation. John's altimeter reads 1,400 meters. "Guans live above 1,900 meters, so be prepared for a tough climb."

I must say this took me aback. It's like climbing the Euromast about nine times without stairs or railings, through a pitch-dark rainforest.

Ever since I first saw the Horned Guan in a book as a child, I've wanted nothing more than to see it in the wild—this huge, black-and-white, gallinaceous animal with pink legs and a velvety black head. The unique bright red horn on his forehead stands straight up, like a raised thumb. These birds can be found only in the remote mountain forests of Guatemala and southern Mexico. Few people have ever seen one in real life. However, in this inhospitable area I am mainly concerned with surviving.

John and the Guatemalan guide are climbing at a murderous pace. I focus on John's heels, trying to fit my feet exactly into his footprints. I jump over tree roots, climb over fallen trees, and avoid overhanging branches. In spite of being in good condition, I am forced to take a break twice. Resting my hands on my knees, I gasp for air. An hour before dawn, at an altitude of 1,935 meters, I am soaking wet with sweat down to my underpants, completely exhausted.

John points at a pile of dark gunk full of fig seeds.

"Fresh poo from the Horned Guan; we're right on the money."

And now the waiting begins. I forgot to account for the temperature, which is near freezing at this altitude, and soon my teeth are chattering. I long for the first rays of sunlight, but at the moment I see only the beams of light from our headlamps illuminating the ghostly forest. A lava-spewing volcano produces a bright orange glow on the distant horizon. Finally the sun comes up, and I get a bit warmer, shuffling along in the sun's rays that reach the undergrowth through the dense canopy. We stare tensely at the treetops through our binoculars, looking to spot any movement. A few times we think we hear a guan, but it turns out to be a false alarm every single time.

John says, "If he doesn't show up in the first light of day, we can forget about it."

I look at my cell phone and see that it is almost eight o'clock. I curse under my breath. I can already imagine returning to the lodge empty-handed.

It didn't work out, Dad. I'm a worthless bird-watcher.

It is past nine o'clock and still no trace of a guan. We split up to increase our chances, but with each passing minute my heart sinks deeper. This isn't going to happen to me, or is it?

Suddenly, I hear a whistle from somewhere higher up the mountain. This can mean only one thing, and I sprint uphill. When I finally reach our Guatemalan guide, I'm coughing my lungs out. I look up at the treetops filled with hope, but the man shakes his head and nods toward the deep ravine to the left of the trail. "Follow me."

We climb down the slope as fast as we can. Branches hit my face, and I keep slipping. The guide moves like a mountain goat, and it is difficult for me to keep up with him. A rock wall blocks our passage. Before I know it, the guide jumps a couple of yards down. All I can think about is the guan's red horn, and I leap after the guide like a madman, making a rough landing on the rocky soil. I quickly check my ankles, making sure I haven't broken anything. We sprint farther down. About a hundred yards farther, the guide stops abruptly, and only a bush prevents me from bumping into him at full speed.

"Over there!"

I see a huge fig tree. My hands tremble when I aim my binoculars at the dark figure half hidden in the canopy. There it is, just as I saw it sixteen years ago in that picture book: the Horned Guan. I breathlessly watch the prehistoric-looking bird, which in turn stares back at its perplexed observer from a thick mossy branch. After fifteen minutes, during which I remove my binoculars from my eyes only to take pictures, he opens his wings and disappears from view for good.

Scheveningen, Netherlands (1986)

Where did this bird madness come from? According to my mother, it started from when I was born. When she walked me around in a pram, I always used to look up at the canopy eyes wide open.

"He is such a quiet and observant kid, I'm sure he's going to be a professor later," she told her friends.

My parents gave me a black-and-white cuddly toy, which, when I started babbling my first words, I named Pica pica. For my seventh birthday, my grandmother gave me *Seeing Is Knowing!*, an old-fashioned bird book that depicts all the birds that can be observed in the Netherlands. In addition to the Dutch names, this small book also provided the Latin names for all the birds, and this is how I discovered that *Pica pica* is Latin for magpie. My mother's prediction seemed to come true. However, many years later, when I failed my master's thesis in archaeology for the third time in a row, it became painfully clear that I wasn't cut out for science.

Every child is a nature lover, I am thoroughly convinced of that. And it doesn't take much to fuel this love. In my case, a small scoop net, a magnifying glass, and a bucket were sufficient. As a young boy, I used my scoop net to empty the neighbor's pond, making drawings of the water beetles, salamanders, and frogs I captured and made swim around in my mother's vases at home.

Most of all, I loved the beach, which is a good thing growing up in Scheveningen. Like my friends Willem, Titiaan, and Maurits, I was the proud owner of a push net for shrimp, which we used on most summer days for catching shrimp, sand crabs, and plaice, which we then released in self-made basins made from sand. With each rising tide, our temporary aquarium was taken back by the sea, despite our efforts to reinforce the dikes with sand and shells.

There was great excitement when one of us caught a pipefish in his net, a sinuous, thin fish that somewhat resembles a long-snouted eel. We were completely captivated by this strange animal, and we pulled our parents away from their beach chairs to come and admire our catch.

At that time I already understood that nature is never boring. If you pay attention, you will always see, hear, smell, or feel something surprising, whether you are walking around in a tropical rainforest or in your own backyard.

On a hot summer day in 1995, Titiaan, Willem, and I found a cormorant on the beach. The bird sat on the high-tide line between the seaweed and washed-up debris, suffering from the blazing sun. Until then I had seen cormorants only from a long distance, and then they appeared mostly black, but now I noticed how beautiful its plumage was. His tail and mantle feathers had a sort of oily green sheen, and the distinctive dark feather edges gave him a scaly appearance, like a snake's skin. That, in combination with his turquoise-blue eyes, made him look somewhat reptilian. He was an adult bird in breeding plumage. Above his legs was a prominent white spot, and his crown and neck were covered with fine, snow-white feathers.

With his mouth open and his eyes half closed, he allowed us to approach him within a yard. I immediately realized that something was wrong, otherwise this shy waterbird would have flown away well before we could come close. We had to catch him and take him to the bird shelter. There was one big problem: How were we going to transport this big bird to the other side of town on our bicycles? Fortunately, my mother was lying on her towel about a hundred yards away, unsuspectingly enjoying her day off. And she had come to the beach in her brand-new Renault Twingo.

"In my new car?" my mother asked a moment later.

It took some convincing, but eventually she gave in. I caught the cormorant with a towel and put him in a cardboard box that Titiaan had found at a beach club. Together we left in my mother's car for the bird shelter. After the first 200 yards, the cormorant raised his cloaca above the rim of the box. A thick stream of gray-white gunk sprayed across the back seat, filling the car with a suffocating fishy smell.

I carefully carried the box with the cormorant into the bird shelter.

"He doesn't have long to live," muttered the vet. She carefully lifted the bird from the box and placed him on the table. His head fell down, and his eyes were closed.

"He probably swallowed a fishing hook, which caused him to bleed internally . . ."

In the middle of her sentence, the cormorant cramped up. He lifted his head backward and released his last breath with a hissing sound. The three of us, just little boys, stood around the treatment table, feeling completely defeated.

On the way back home, we stared ahead in silence. Through the open roof, I saw a flock of cormorants flying overhead. Titiaan and Willem followed my gaze and saw the V formation just disappearing behind the trees.

"They're like jet pilots flying a lap of honor for their dead friend," Titiaan said.

He couldn't have said it better.

Scheveningen, Netherlands (2016)

Camilla, my girlfriend, says it's nerves. The 366-day* journey through forty countries, which starts today, New Year's Day, has become too much for my usually pretty-strong stomach. Last night, I had to run to the bathroom every half hour. In the early morning, I walk up to the roof terrace for a breath of fresh air. Meanwhile, the fireworks lovers hanging out in Scheveningen have shot all their rockets. Only an occasional stray bang can still be heard from afar, from the center of The Hague. It really and truly is January now. The smoke from the fireworks hangs still in the air and tickles my nostrils. I always hated fireworks, and knowing that tens of millions of euros worth of pollution and noise have just been shot into the air over the Netherlands in just a matter of a few hours only makes my aversion greater. At that moment, a familiar sound arises from one of the nearby backyards. I strain my ears and hear the unmistakable jubilant song of a Common Blackbird. It's as if this heroic bird wants to show that he has not been deterred by the deafening bangs and flashes of light.

This is the kickoff: Bird species number one is now a fact. 6,042 to go this year to set a new world record.

The ringing of the doorbell is the official start of my Big Year, a whole year of birding. It's New Year's Day, six o'clock in the morning, and Max is standing at the front door with the car.

"Are you ready?" he asks Camilla and me, blowing in his hands.

Max will travel with me for the next two and a half months through Asia and New Guinea. We know each other from back in the day when we went to birding camp. I am two years older, and I considered him to be a bit like my younger brother in birding. During his high school years, he temporarily lost his interest in birding. Other hobbies, such as parties and girlfriends, were given priority. I went to study in Groningen, and Max disappeared from view for a long time. A few years ago, he unexpectedly called me to ask if I wanted to go

---

* I deliberately chose a leap year since it gave me an extra day.

birding again. And while we stood along the shore of the Lake IJssel-meer looking at large rafts of Tufted Ducks and scaups, I talked about my Big Year.

"Is it okay if I travel with you for a while?" he asked. "I could take pictures for you to use in your lectures." He had been particularly fanatical about bird photography lately, and he had always wanted to go and see Southeast Asia.

I had to think about it for a while—we hadn't spoken in years after all—but after lettting it sink in for a couple of days I finally agreed. We could share the cost of the guides and accommodations and, more important, it would be a lot more fun than traveling on my own.

Many birders across the globe were apprehensive of my plan to start my Big Year in the Netherlands. Wouldn't it be much better to start somewhere in the rainforest of South America or in Australia? Yet this had been a conscious choice: During the winter months, the Dutch delta accommodates huge concentrations of many different species of geese and ducks. Our country is a great refuge for wintering waders and waterbirds, and I wanted to show this to the rest of the world.

We have a tight schedule that will allow us to see more than a hundred species today, at least in theory. The plan is to make a big tour across the delta, and to drive directly from there to Schiphol.

I'm holding my breath, as it has rained almost continuously for the past two weeks, and every birder knows that birds tend to remain hidden in the pouring rain. But when we arrive at the first birding site at sunrise, there is not a cloud in the sky and no wind at all.

The air is fresh and salty, and I take a deep breath. My nausea from the previous night is gone. The rising sun sets the sky ablaze, and we hear the calls of wigeons, curlews, and Brant Geese all around us. How beautiful nature can be in the Netherlands!

Every species seems to be cooperating today, and by evening we already count 117. However, Camilla's stomach is not cooperating. It's as if she has caught my nausea. Her anxiety increases as we get closer to my departure. While we drive to Schiphol, I put my arm around her. "I'm going to miss you so much," she whispers as she rests her head on my shoulder.

Schiphol, Netherlands (2006)

After finishing high school, I went on a trip. And by this I don't mean backpacking through Thailand for a month—I went birding for seven and a half months. I had done several side jobs for almost a year, as a dishwasher, beer seller at concerts, and assistant chef in a cooking studio. The last job was very short-lived, as I didn't even know how to fry an egg and I drank just one too many beers every night. In the end, I had saved a few thousand euros, and my parents bought me a round-the-world ticket as a reward for my high school degree.

And so it happened that, on January 3, 2006, at the age of nineteen, I walked on the moving sidewalk at Schiphol on the way to my gate. The raspy voice of Bob Marley sounded through the earplugs of my MP3 player: *Emancipate yourself from mental slavery. / None but ourselves can free our mind.*

I felt as free as a bird while adventure beckoned on the other side of the world.

During that trip I did nothing but birding, from sunrise to sunset. Sabah, the Malay part of Borneo, made an especially deep impression on me. In addition to large numbers of hornbills and other enthralling bird species, such as the Bornean Ground Cuckoo and the Malaysian Rail-babbler, I saw an orangutan for the first time in my life.

Everything radiated beauty, but I also witnessed the consequences of the rubber and palm oil industry. As I traveled by bus from the west to the east coast, I passed seemingly endless rolling hills with palm plantations and logging plains. Occasionally, I could see a huge tree stump standing out among the commercial plantations, it being the only remnant of a 200-foot-tall forest giant. Less than twenty years ago this was all tropical rainforest, with hornbills, orangutans, and even the nearly extinct Sumatran rhinoceros. As I stared out the window and watched this apocalyptic setting pass by me, I wondered if there would be any forest left in ten years' time.

After a three-month journey through Malaysia, Australia, and New Zealand, I arrived in Santiago, the capital of Chile. From there I hitchhiked over the Pan-American Highway to Peru, a journey of nearly 1,600 miles. At nightfall I pitched my tent along the road, in the middle of a bone-dry moonscape. At night it was very quiet, except for a single stray truck. No insects, no wind, no birds. Nothing. I lay on my back on a mat, and I stared at the endless starry sky through the open zip of the tent canvas; the silence was so deafening that I could hear my own heartbeat.

Peru is a true El Dorado for birders. The enormous diversity of landscapes—from the bone-dry Atacama Desert to the snow-capped peaks of the Andes Mountains and the brooding jungles of the Amazon rainforest—are home to more than 1,900 bird species, more than twice as many as in the entirety of Europe.

I ended up staying there for four and a half months. Like the locals, I rode in overcrowded buses and hitchhiked on truck roofs to the most remote corners of the country. Every area I visited brought a series of new bird species.

On my way from the Amazonia lowlands to Cuzco, a city at 3,300 meters altitude, I sat on the roof of a truck that climbed painfully slowly up a dirt road. My legs and buttocks felt sore from constantly sliding over the wooden planks that covered the roof. I shared company with a few Peruvians who were on their way to the city with their merchandise. Like them, I held on tightly to the metal railing so as not to fall off the truck in case of unexpected bumps in the road. To our left, we could see a steep, wooded slope, and to the right a ravine of several hundred yards in depth. We passed a burnt-out bus that had missed the road and ended up in the rocks far below us—a silent witness to a drama that had taken place here. One wrong steering movement by our driver and we, too, would end up down there. I was quite nervous, as I had already watched at least three empty half-liter cans of beer flying out of the open window of the driver's cabin.

Sitting next to me was Rob, a Dutch birder who had been living in Peru for two years. We had just spent a week birding together at a research station in the Tambopata Nature Reserve. We calculated that we would arrive in Cuzco sometime in the evening, but two flat tires, a roadblock, and an average speed of 15 miles per hour meant that we were only halfway there by sunset. Around midnight, we finally reached the mountain pass at 4,000 meters. The temperature was well below freezing, and Rob and I still sat on the exposed roof, shivering with cold.

"Would James Clements have experienced anything like this in his Big Year?"

I looked at him questioningly. "Who is James Clements, and what is a Big Year?"

Rob told me about the American ornithologist who set the world record for birding in 1989: "He traveled around the world for a year, visited all continents except for Antarctica, and watched 3,662 bird species. No one has ever done this after him."

A Big Year . . . those words had a magical ring to them. Birding was my great passion, so what could be more beautiful than nonstop birding around the world for 365 days? Countless species that, so far, I had seen only in bird books; new cultures, languages, and tastes; and the most beautiful natural areas in the world. A whole year long. If this American could do it, I could do it! This could become the great adventure that would put me on the map as a birder. And there, in the dead of night, blue-lipped from the cold in the Peruvian Andes, I decided that someday I would become the Big Year world record holder.

Schiphol, Netherlands (2016)

I can see how hard it is for my mother to say goodbye to me.

"I'll be careful, Mom, I promise."

My father puts his arms around me.

"Whatever happens, always hold on to the railing."

I hug them extra tight. Without my parents, this whole adventure would not have been possible, I'm fully aware of that. They have always supported me in my hobby, through thick and thin and unconditionally. Birding was the main theme of our family vacations, even though they weren't birders themselves. If I came home an hour late for dinner because I had been listening to a nightingale singing in the dunes, it wasn't an issue. In fact, the following evening my father would join me. When I came home from high school as a thirteen-year-old, deeply distressed because I was being bullied for my hobby, my mother said, "Ignore them, Arjan. They're only jealous." After which she put on her coat before heading to the front door. "Come on, grab your binoculars. We're going to the Oostvaardersplassen Nature Reserve."

After checking in for my flight, I hug my parents one more time and plant a kiss on Camilla's forehead. I say that I will see her again during my stopover in the Netherlands on April 1. Then I get a lump in my throat and bite my lip to hold back my tears, as it will take three months, ten countries, and forty flights before I will see her again, and then only fleetingly. I consider myself lucky to have found someone with the same love for birds, who understands better than anyone else that I have to do this, and who respects and appreciates me for it. I realize all too well what I'm risking with this adventure: A year is a long time, during which so many things can go wrong.

As Max and I walk to the boarding gate, all kinds of thoughts race through my head. Will I remain healthy throughout the year? Will all my flights run smoothly? Will my relationship last, or will it be too much for Camilla? Am I going to break the record? But then I put on my headphones and, just like ten years ago, I'm listening to the voice of Bob Marley: *Won't you help to sing / these songs of freedom!*

I close my eyes and take a deep breath. My Big Year is starting for real now.

Scheveningen, Netherlands (2014)

One day in September, I had dinner with my parents.

The idea of a Big Year had been whizzing through my mind ever since, on the roof of that truck in Peru, Rob told me about Clements's record. In 2008 a British couple, Ruth Miller and Alan Davies, did a world Big Year, and I had followed their progress closely. At the end of their year, they set the new world record at 4,341 species. This made me only more motivated to undertake such a Big Year myself someday.

If I was going to do it, it had to be done quickly. If I waited a few more years, such an endeavor might be impossible due to circumstances, such as a family or a job. It was kind of now or never.

While my mother put the food on the table, I casually disclosed my plans: how Clements, in 1989, observed 3,662 bird species in the span of one year; how Ruth and Alan broke that record; and how I wanted to use my attempt to break this record, write a book, and gain recognition in the birding world, so that I would eventually be able to turn my hobby into my job.

"You should see it as a kind of investment."

With that sentence I concluded my sales pitch.

"Traveling for a year—that's a very long time," uttered my mother. "But if this is your dream, then we won't stop you."

"I could do the back office from home," my father said enthusiastically. He had retired a year earlier, and it was clear that he had found a project to get into.

I looked at them perplexedly. "So you think it's a good idea?"

My parents nodded.

This was unbelievable, I expected that they would call me crazy, but the opposite turned out to be the case. I should have known: When it came to my love for birds, my parents were always there for me.

That same evening I posted a tweet sharing my plans. I hadn't the slightest idea how I was going to tackle anything, specifically, but at least the first step had been taken. That is the most important thing with such big plans—you have to make a decision and then go for it. Otherwise, when you are eighty years old, you might ask yourself: How would my life have turned out if I had taken that leap of faith?

Banksy, a famous street artist, once wrote on a wall: "You're always one step away from changing your life completely."

Yet that night, I lay wide awake in bed. What on earth was I doing?

The next morning, it appeared that my tweet had gone viral in the birding world. Someone from the Bird Information Center on the island of Texel had left a message on my voicemail. They would like to sponsor me on behalf of Swarovski Optik.

I was overexcited and replayed the message on speakerphone so that my parents could hear it, too. Swarovski is the absolute state of the art when it comes to binoculars and telescopes!

While I was still euphoric about my very first sponsor, I received a text message from a birder friend with a screenshot of a newspaper article.

AMERICAN TRIES TO SEE 5,000 BIRD SPECIES IN ONE YEAR, the headline read.

I shook my head and quickly read the rest of the article. Thirty-year-old Noah Strycker was going to do exactly what I would try to do the following year. The piece covered the nonstop journey that would take him through forty-one different countries. I quickly added up the numbers: 1,500 species in Asia, 500 in Oceania, 1,000 in Africa and Europe, and another 2,000 in South, Central, and North America.

There were two options left to me: I could throw in the towel now, or I could go a step further. I decided to go for option two.

Dubai, United Arab Emirates (2016)

"Good morning, sir!"

Those sleeping pills had worked better than expected. According to Max, I'd kept him awake for at least four hours with my snoring. He pushes up the shutter of his window. For a moment we are blinded by the bright sunlight reflecting off the wing of the plane, but then we slowly see the contours of the Burj Khalifa, the tallest building in the world. We are arriving in the Las Vegas of the Middle East.

In the arrivals hall we are met by Mike, a British birder who works as a pilot and has lived in Dubai for over ten years. His bald head is tanned by the ever-scorching desert sun, and his eyes are hidden behind sunglasses. Binoculars hang from his neck, and a camera with a gigantic telephoto lens dangles from his shoulder.

Dubai is littered with skyscrapers, and we regularly pass by huge construction sites, the starts of the next multimillion-dollar projects. A considerable part of the city's population consists of poor foreign immigrants, brought here to work hard for a pittance, twelve hours a day, six days a week, in the sweltering heat. The dark side of all the splendor.

Here, nature will always be subordinate to spatial development. A tidal zone can turn into a construction site overnight when a billionaire has decided to build an apartment complex. But a dry stretch of desert can also be transformed into a green oasis just like that, because that same billionaire wants to build a golf course. Mike says that birders have to be flexible here. It's a matter of making do with what you have.

In the middle of a busy roundabout, he suddenly slams on the brakes and points out the window. "Look!"

A White-tailed Lapwing casually walks on the short-mown grass, in between sprinklers. I expected that we would have to put in considerably more effort to catch this elegant, brown-gray wader with long, bright yellow legs. It normally breeds on the steppe of Kazakhstan and Uzbekistan and winters in wetlands and along rivers in northwest India and Pakistan. All around me I see concrete, stone, and sand. The grass of the roundabout is the only green in the area. The lapwing probably lost course during its migration south and decided that this short-trimmed patch of grass was the best place to spend the winter.

———

That evening we stay with a friend from my college days who recently started living and working in Dubai. He took the next day off to take us in tow.

We get into the car just after five o'clock in the morning. It is still pitch dark. There is hardly anyone on the road except for us. We soon leave the skyscrapers and construction sites behind us, and a little later we drive straight through the desert while the rising sun colors the hundreds-of-feet-tall sand dunes on the horizon. The image reminds me of that famous sunrise in *Lawrence of Arabia*.

Farther east, the immense sandpit turns into rocky slopes with sparse vegetation. We stop at Wadi Masafi, a dry riverbed that quickly turns into a swirling mudslide when there's heavy rainfall. Here we are looking for the Plain Leaf Warbler, a brown-gray warbler of less than 3.5 inches, which winters relatively locally near the Persian Gulf. A narrow path leads through steep, rocky slopes of reddish-brown rock. Except for the sound of our footsteps, it is dead quiet. What a contrast to the crazy city of Dubai.

Colombo, Sri Lanka

When putting together my travel schedule, I had booked as many night flights as I could. This has two advantages: I lose as few precious hours of daylight as possible, and each one saves the costs of an overnight stay. The downside is that Max and I arrive in Colombo at 2 a.m. We've had barely any sleep and are jet-lagged.

Our guide is already waiting for us in the arrivals hall. We take a van to Sinharaja Forest Reserve, the place with the highest concentration of endemics—bird species that are found nowhere else in the world. During the trip, I try to catch up on some sleep, sliding across the back seat, using a rolled-up sweater as a pillow. This pillow proves to be useless. Every time I nod off, we pass through a hairpin turn, causing my head to bump against the door of the van.

We arrive at the entrance of the park just before sunrise. The timing couldn't be better. Seeing the boundless rainforest-covered hills instantly makes me feel good, because somewhere in that impenetrable greenery awaits the Serendib Scops Owl, one of the greatest ornithological discoveries of recent decades.

In the mid-1990s, ornithologist Deepal Warakagoda heard an unknown noise at night in the wet rainforests of Kitulgala. He was unable to determine the source of the sound. Every time he heard it, his suspicion grew that it was an undiscovered species of owl. After six long, unsuccessful years of being on the lookout, he finally caught sight of the bird, and for the first time since 1868 a new species for science was described from Sri Lanka. The entire population probably consists of no more than a few hundred birds.

The discovery of the Serendib Scops Owl shows how little we really know. The rainforests are true treasure troves that have not divulged all their secrets yet. The only way to unravel them is through conservation and research. But the logging industry is merciless, and it is a race against time. Every second, an area of rainforest about the size of a football field is being felled somewhere in the world. This is equivalent to the surface area of New York City every day or the surface area of Italy every year.

⁓–⌣–⁓

The Sinharaja forest is virtually untouched, with trees over a hundred feet tall, each of the forest giants many hundreds of years old. One tree next to the path is completely covered and swallowed by a strangler fig. That fig started hundreds of years ago as a seed in bird droppings on the branch of its host. All that remains of its victim now is a rotten tree trunk. In the rainforest, trees fight a continuous battle: Only the strongest will eventually settle with their crowns in the canopy.

It is quiet in the forest, and at first we don't see a single bird. Birds in tropical rainforests usually live in a mixed species flock, sometimes containing dozens of species that work closely together. This collaboration is complex. Some species stay only at the top of the treetops and look for berries and fruit there. Just a little lower are the insectivores, ready to catch any unfortunate beetle or spider chased from its hiding place by a fruit-eater. There are species on the forest floor that pick up leftovers from the ground. When danger threatens and a bird sounds the alarm, they all take to their wings. In short, the mixed species flock operates as an organized body in which each participant has its own specialist role.

Scientists determined that such flocks in Sinharaja are among the largest and most complex in the world. But first, it's a matter of finding such a flock.

⁓–⌣–⁓

In the distance we hear the magpie-like call of an Orange-billed Babbler. The guide gestures for us to follow him. As we make our way through the dense vegetation, the sound of birds slowly swells, and before we know it we find ourselves in the middle of a mixed flock. Left, right, high in the trees and low to the ground, we are surrounded by birds. We see a Green-billed Coucal, with its light green, downward-curved beak. Two yards higher, we see a striking green and white bird with a long tail and a bright red mask of short bushy feathers—a Red-faced Malkoha! It all goes very fast, a couple of minutes tops, and when the flock is gone, we have seen almost half of the endemics of the park. This is a perfect example of why birding is such a great hobby: One sighting can turn even the dullest day into a day to remember. When I walk around in Vlieland, in the Netherlands, in October, looking for North American or Siberian vagrants, one thought keeps me going, even though I've been looking at the same flocks of tits and Goldcrests for weeks on end: There will come a day when I suddenly bump into a new bird for the Netherlands.

We explore the forest tracks of Sinharaja well into the afternoon, where we can add a few new species with each flock. When we arrive back at the entrance of the park, tired but satisfied, our guide turns out to have one last surprise in store. He crawls into the thick undergrowth and we follow. He stops a few yards away and points to a dark tangle of ivy: "Look over there, Serendib Scops Owl!"

And there, between the bamboo and the ivy, we see a sandy-colored owl with black spots on his chest and a slightly brighter V on his face. He stares at us with one eye half open. He is barely bigger than a European Starling, and his camouflage colors keep him from standing out among the dense foliage. I now understand why it took Dr. Warakagoda so long to lay eyes on him.

Scheveningen, Netherlands (2014)

From the moment I took the plunge, it took me about fifteen months to plan my Big Year. This may sound like a long time, but there was a lot to be done. My financial situation was an important factor. I had just finished my archaeology studies and was working as a bartender in Amsterdam, which is not exactly a solid foundation for funding nonstop travel for a year. Fortunately, my parents had promised that I could borrow money from them in case of an emergency, and I had my apartment in Scheveningen that I could rent out during the trip.

I would spend quite a few hours in airplanes during my year, and that bothered me, as flying is extremely bad for the environment. Like Noah Strycker, I would follow a carbon-neutral program to offset my carbon footprint. I paid a little extra for each flight; that money would be used to give households in developing countries access to sustainable energy. However, I knew that this alone would not be enough to make up for a year of travel.

"If you are really going to do this, make sure that this journey will allow you to contribute to a worldwide awareness process," my father said.

With these words, he steered me in the right direction.

~——•——~

About one-eighth of the more than 10,900 bird species on our planet are threatened with extinction. The situation is critical for more than 200 species; if conservation measures are not taken immediately, they will become extinct in the near future. There are hundreds of examples where protection came too late (or never came at all). The best-known example is, of course, the Dodo, a species of giant flightless pigeon that was wiped out by humans within two centuries of its discovery on Mauritius.

I had decided that my trip would raise money for the most endangered bird species in the world. I heard about the Preventing Extinctions Program (PEP), an innovative program created by BirdLife International with the purpose of saving the most endangered bird species from extinction. PEP has appointed species guardians all over the world, people and organizations who write down, coordinate, and ultimately implement conservation measures. Often these species guardians have been working with certain species for decades, and, therefore, know better than anyone what it will take to save it. Countless bird species have already been saved this way.

"I want to raise half a million euros," I said. I've always been an incorrigible optimist.

Fortunately, my father was more realistic. "Collecting money for a good cause is not easy, and half a million euros is a lot of money. I would make it a hundred thousand euros. You will see that raising this kind of money will already be quite a challenge."

In retrospect, I am grateful to him for this adjustment.

I sent an email to the director of PEP, writing to him about my plan to raise money for them for a year.

"Dear Arjan, you may now call yourself Species Champion for the BirdLife Preventing Extinctions Programme," he wrote back.

———•———

I managed to make regular radio and television appearances to promote my upcoming world record attempt and charity. Pure bluff, because at that time I had nothing at all. No flight had been booked yet, and I had only a vague idea of what my itinerary would look like. I planned to travel large parts by public transport and without the help of guides. After all, I had always done it this way during my birding trips. But meanwhile, Noah had been on the road for a few months, and every day I saw his species count skyrocket on the site of Audubon—the American bird protection organization. He was already onto more than 2,000 species, and he still had four continents to go. He had the best guides everywhere he went, allowing him to score a maximum number of species in a relatively short period of time in every country. It was clear that he was going to set an almost unbreakable world record. Therefore, I needed to fine-tune my plan.

I thoroughly studied Noah's itinerary and tried to find where I could make improvements. I swiftly sent hundreds of emails to guides, tour companies, and lodges all over the world, asking if they might be interested in helping me. My campaign was successful: Most people responded with heartwarming enthusiasm, and by mid-2015 I had built a network of guides who would be available to me. Moreover, dozens of Dutch birders had asked me if they could join me for a little while; they were more than happy to give up their vacation plans to join in with their compatriot's record attempt. Strangely enough, I had not expected any less: The Dutch birding community is a close-knit group. We meet up a few times a year during birding weekends on the Wadden Islands. Everyone knows one another, and if someone is working on a cool project, it doesn't take long to find help and supporters. We go to the pub together, jointly celebrate birthdays, and we like to travel together, like one big family.

Meanwhile, my radio and television appearances also paid off, and sponsors started to trickle in. Slowly but surely, my Big Year started to take shape and substance. I would visit forty countries and every continent except for Antarctica—Noah had started there and was able to observe only a few dozen species in eight days. I had calculated that if all went well, I could possibly observe 6,000 species in one year.

Noah's list followed the Clements counting rule. Former world record holder James Clements was the founder of this taxonomic classification, which mainly looks at DNA to determine whether a bird should be considered a separate species or not. In Europe, we follow the taxonomic classification of the International Ornithological Committee (IOC). The IOC classification looks at a combination of bird song, appearance, morphology, geographical distribution, and DNA to determine whether a bird deserves species status. The IOC is slightly more progressive, which means that a few hundred more species are included, according to this classification. Therefore, in order to break Noah's record in accordance with the IOC classification, I would have to observe 6,119 species instead of 6,042.

New Delhi, India (2016)

People are sleeping around smoldering fires on the sidewalk and on the median strip. Cows with humps hanging limply on their backs defecate in the middle of the street, while their owner squats on the roadside a few yards away. Traffic rules don't seem to apply here, except for the law of the fittest and the most reckless. Our driver, a tall Indian man named Rakesh, navigates his 4×4 through the chaos at a dizzying speed. The scar on his face makes him look like a bad guy in an Indiana Jones movie. Every time he nearly runs over a pedestrian, he glances in the rearview mirror and smiles proudly, showing off his two gold teeth.

We come to a screeching halt in front of a dilapidated hotel with peeling paint and boarded up windows. Sander, our travel companion for the next three weeks, is awaiting us at the door.

I know Sander from Groningen, where we lived in the same student house for five years. He is a tall, curly-haired young man who has an unrivaled talent for throwing parties. His birding career started, not unsurprisingly, with a party. After a good night out, he asked me if he could join me to go birding in the Lauwersmeer. His hangover melted away like a snowflake at the moment he laid eyes on his first White-tailed Eagle. That same day, he went out to buy a pair of binoculars, and he has been hooked, just like me, ever since.

—

Miraculously, we arrive unscathed at our small hotel near the Keoladeo National Park. Finally, I will have a chance to get some sleep. At least that's my intention. The tandoori chicken we had at a dingy roadside restaurant along the way is playing up. A sharp cramp in my stomach is immediately followed by a wave of nausea. I have food poisoning. I run to the toilet; I have severe diarrhea and need to throw up. When I walk out of the bathroom a little later, I sense that someone is looking at me from the bar.

A man with a concerned look asks me, "You got the Delhi Belly?"

I manage to squeeze out a smile and raise my thumb in affirmation.

"Here," he says, pouring a glass of whiskey. "This will help."

The man claims to have a foolproof cure for food poisoning: a large glass of whiskey (or any other spirit for that matter), a few cups of hot tea, two Tylenols, and a long night's sleep.

"The next day, drink plenty of water and eat bananas, and you'll be back on your feet," he says, giving me an encouraging nod.

I'm prepared to do anything to feel better, so I pinch my nose and empty the whiskey glass in one go.

—

In the mid-eighteenth century, Maharaja Suraj Mal, the erstwhile ruler of the state of Bharatpur, had built a dam southwest of the city of Bharatpur. The lower natural area around the Keoladeo temple flooded with water, and the resulting lake became part of the Maharaja's hunting grounds. At times, he and his court held large-scale hunting parties where thousands of ducks were shot in one single day, which was a normal thing to do in those days. About forty years ago, Keoladeo was designated as a national park. Hunting was no longer allowed, and the area became a paradise for waterfowl and birders.

When the alarm goes off, I feel as weak as a rag doll, but I'm lucky: Today we will be birding in a bicycle rickshaw, a kind of moving armchair. So I will be able to watch birds and have a rest at the same time. We drive at walking pace along the many lakes in this otherwise fairly dry area. Meanwhile, I drink tons of water and eat one banana after another.

I still feel sick and hang half out of the rickshaw. Suddenly, two black-and-white songbirds with long tails and a light peach chest fly over my head and land on top of a tree. I jump up and grab my binoculars. To my surprise, I see a pair of White-bellied Minivets. Normally, one has to travel far inland to Rajasthan to have a chance to see this nomadic species. Here, it can be seen only once or twice a year. I jump out of the rickshaw and run closer to take a photo. My nausea has magically disappeared. What seeing a rare bird can do! Or maybe it was the whiskey that cured me. We will never know.

Guwahati, India

In mid-January we arrive at Guwahati airport. Sander, Max, and I are joined by another friend from my student days, Arnoud. Our fifteen-day expedition takes us through the Northeast India provinces of Assam, Nagaland, and Arunachal Pradesh. The last state is claimed by both India and China. At this moment, India is in charge, but the many army bases bear witness to the tense situation.

Peter and Rofik, two of India's best bird guides and the region's experts, lead the expedition. On the way to our first destination, they tell tall tales of treks through the Himalayas and encounters with big cats. When I doze off in the back seat, I dream about tigers and snow leopards.

"Wake up!" Peter taps me on the shoulder. "We have to show our passports."

At the military checkpoint we are treated with suspicion; they hardly ever see tourists here. We are not allowed to film and have tucked our binoculars and cameras deep in our backpacks. After taking a look in the trunk and a thorough inspection of our papers, they allow us to drive off.

The road climbs up parallel to a fast-flowing river. On both sides, there are steep slopes covered with subtropical mountain forest. As we get higher, the road gets worse, and we see fewer and fewer other cars or people. We pass villages consisting of a few rickety wooden houses. Dogs run alongside the car, barking loudly, biting at the hubcaps of our 4×4 while risking their own lives.

After almost half a day of driving, we reach our base camp for the next two nights, at an altitude of more than 2,500 meters.

Set up in the middle of nowhere, Lama Camp consists of a few canvas tents and a small wooden dining room with no electricity or running water. You can hardly get much farther away from civilization. This camp is well known among birders because the Bugun Liocichla was discovered here in 1995. Although it was not until eleven years later that this bird was defined as a new species, it was immediately awarded BirdLife's highest possible conservation status, "critically endangered"—in danger of becoming extinct in the very near future. When defining a new species, usually a number of specimens are "collected" as museum specimens, but this was forgone with the Bugun Liocichla, as its population is estimated at merely a few dozen birds.

That afternoon we start our search. We walk for miles up and down the road, over and over again. This is the best tactic, because the farther you walk, the more chance you have of encountering a Bugun Liocichla. However, by evening we have only heard it from a great distance.

We eat our dinner by candlelight and sleep in a tent. It is extremely cold, even with a thick sleeping bag and an extra blanket. We have to stay dressed, including our hats, to keep warm. I'm enjoying it. This may sound strange, but it's exactly how I envisioned my Big Year: one big adventure. And these kinds of hardships form part of that.

We resume our search in the morning, but the Bugun Liocichla refuses to cooperate. Again we hear two birds calling, now a few dozen yards farther down in the valley, but the dense undergrowth impedes us from seeing them. This is our last chance, as our tight schedule does not allow us to stay here any longer.

Nearly everyone who visits this place has a chance to see this bird, except for us! We never found it. I feel almost as sick as a few days ago but not because of the food. The first big miss of my Big Year is now a fact, even though I've heard the Bugun Liocichla and I can officially count it for my world record attempt. Noah was the first to do that, and if I want to keep up with him I have to do the same. In a way, I can understand it: The sound of a bird often divulges just as much as its appearance, but that does not alter the fact that I would have loved to have seen it.

———•———

Past Lama Camp, the road winds farther up to the Eagle's Nest Wildlife Sanctuary. It is situated in one of the largest contiguous subtropical forests in Asia. Leaving the Eagle's Nest Pass behind us, we enter the park. We see green slopes shrouded in mist all around us. Every now and then, a bird of prey circles up from the valley.

There is only one road through the park, from the pass at an altitude of almost 3,000 meters, all the way into the lowlands. This road leads through a completely virgin forest, except for a few tiny settlements. This is one of the few places in the world where nature lovers can experience a contiguous ecosystem with such a difference in altitude, thereby offering an exceptional spectrum of plant and animal species.

We see dozens of new bird species, including the Fire-tailed Myzornis, a bilious green songbird with a black mask and an unmistakable bright red tail. So far, everything seems to be going well for us. On the third day, however, things start to go wrong. During the night we are woken up by the sound of rain pouring on our tent canvas, and when we drink our chai at five in the morning, it still comes pouring down from the sky.

"If there's rain here," Peter says, looking at the dark sky with a concerned expression, "there could be snow higher up."

And indeed, when we drive up to the Eagle's Nest Pass in our 4×4s, our fear turns into reality. Slowly the rain turns into snow, and as we get higher, the road becomes less and less passable.

"I have never seen this much snow here in twenty-six years," Peter says in disbelief.

I know what this means: We will not see the Himalayan Monal, an insanely beautiful species of pheasant that occurs only around the tree line. But we have an even bigger problem. If our 4×4s fail to cross the mountain pass, we will be stuck here until the weather improves. And that can take a few days . . .

We continue to drive up the road at a walking pace. With every yard we ascend, the snowpack gets thicker and the road becomes more and more impassable. At the steepest parts, we have to get out and push the 4×4s, which—who would have thought?—do not have snow tires. A few times we even have to remove branches, snow, and stones, bare-handed, that have ended up on the road due to avalanches. It looks like we will have to set up camp somewhere in this snow-white no-man's-land.

Miraculously, a pickup truck pulls up behind us. They are farmers from a village in the valley who are on their way to Guwahati with their merchandise. They help us clear the road, and thanks to their combined strength and perseverance we finally reach the pass.

It's only back in the civilized world that we're in range again for Peter to make the necessary phone calls: As expected, the way to the top is completely covered in snow. We have to give up our agenda and leave two days earlier than planned for Kaziranga National Park, in the lowlands of Assam. I try to cheer myself up, contemplating that these kinds of setbacks are part of the process. My predecessors had them, too, and I'm sure there will be more to come during this year. Yet, I feel miserable, and all I can think of is that beautiful pheasant, somewhere up there in an inhospitable world of snow and ice.

Kaziranga National Park, Assam, India

"We have very good chances of seeing a tiger," I said to Sander just before the trip. But I soon realized that the only day we would spend in Kaziranga National Park would never be sufficient. Tigers are rare and shy. A tiger sighting was the main reason for Sander to come along, so I kept my mouth shut—out of necessity—until the plane tickets were booked.

During the car ride, he gradually began to realize that our chances were nil, especially when Peter somewhat condescendingly said: "A tiger? The grass is a dozen feet tall at this time of year. Forget about it."

I keep my fingers crossed that morning as we drive into the park in an open safari 4×4. If we don't see a tiger, Sander won't easily forgive me.

We've been on the road for several hours and, as expected, we haven't seen any tigers yet. We did come across several Indian rhinoceroses. These are slightly smaller than the two African rhinoceros species and have only one small horn on top of their nose. They are sturdily built and have a thick, creased skin, giving them the semblance of wearing armor made of thick plates. In the early 1990s, hunting and habitat destruction reduced their population to less than 2,000 specimens, but intensive conservation efforts by the Indian government ensured that there are now more than 3,500 of these armored tanks walking around.

The fact that we don't see tigers doesn't mean they aren't here, as we see several footprints in the mud on the road.

"A female tiger caught a sambar deer here two weeks ago." The guide points to a bare carcass laying right next to the road.

I notice that Sander seems to feel more and more uncomfortable. He has been having diarrhea almost continuously since his arrival in India, and today is no exception.

"If I don't find somewhere soon, I'll poop in my pants," he says.

He looks around feverishly, but sees nothing other than very tall elephant grass, which may hide a hungry tiger lurking around.

"Stop! I can't hold it any longer!"

Sander jumps out of the 4×4 and runs into the grass. Shortly afterward, we can hear a sigh of relief. A split second later we see him reemerge from the grass, his face as white as a sheet and now and then nervously looking behind him. He climbs back into the 4×4 as fast as he can. "That was the most uncomfortable toilet visit of my life."

———— • ————

"If we don't manage to see a tiger this afternoon, I won't go with you to Nagaland," Sander says during lunch. "I'd rather stay here for a few more days to increase my chances of a sighting."

I cannot but agree with him, because I convinced him to come with me under false pretenses.

When we drive back to the park in the afternoon, I hold my breath. Arnoud and Max have also joined his plan, and I can't bear to think of ending this great journey without them.

We are barely five minutes underway when we hear the barking call of a muntjac. This is the way this small deer species signals imminent danger, and in this park that can mean only one thing: a big cat! The driver parks the 4×4 at the edge of the forest, and the waiting begins. In the middle of the clearing, we see a herd of cows that escaped the watchful eye of their owner and wandered unsuspectingly into the park in search of juicy grass.

"Let's wait," Peter whispers. "I have a good feeling about this."

Again there is the bark of a muntjac. This time it's a lot closer, coming out of the thick undergrowth at the edge of the clearing. We hold our breath. Then Max points to the undergrowth with a bewildered look in his eyes.

"I think I saw something orange with black stripes and . . ."

He hasn't finished his sentence when a huge adult male tiger emerges from the undergrowth. The black-orange stripe pattern sparkles like gold in the morning sun. His mouth is half open, revealing gigantic dagger-shaped fangs. A shiver runs down my spine. What a beast! Bengal tigers can weigh up to 550 pounds, and this beefy specimen will certainly not be far from that.

He has clearly set his sights on the cows. His pace slows down, he slightly lowers his body into creeping mode. He slowly comes closer and closer until he freezes at a distance of about twenty yards, with one front leg still half in the air and his gaze fixed on his prey, like only cats can do. Then he sprints, and before we can blink our eyes he has thrown himself on a calf, which immediately collapses under his enormous weight. The tiger's jaws close around the neck of his prey, which kicks its hind legs a few more times, not standing a chance, and then remains motionless. Everything takes place within a few seconds. The rest of the herd watches suspiciously from a distance. Then the tiger releases the calf and starts another sprint. The cows have clearly lost their survival instinct after thousands of years of domestication. The tiger closes the gap of 200 yards in no time and pounces on an adult cow. It seems to cost him little effort, as if he is pulling a snack out of a vending machine. Once again he puts an end to the life of his prey in seconds with a neck bite that is made with surgical precision, after which we see the tiger dragging the carcass into the tall elephant grass, while an Indian rhinoceros and her offspring are watching this scene at a distance of a mere tens of yards.

We look at each other in amazement. Did we really just see this? Then the euphoria takes over. What an unimaginable sighting! I look at Sander triumphantly. "Didn't I tell you we were going to see a tiger?"

Nagaland, India

Along the road, we regularly see dead birds offered for sale. I see all kinds of species, such as partridges, pigeons, and even a Blyth's Tragopan—an endangered species of pheasant that I have never seen before. A horrible sight. People buy these birds for food or use the feathers as decoration in their homes.

In India's most northeastern province, hunting is a longstanding tradition within local communities. A boy doesn't become a man until he learns to shoot. Historically, they hunted with a bow and arrow, nowadays with a shotgun.

The village where we stay tonight used to be a hunting community, whereas now these people are dedicated conservationists. About ten years ago, Peter set up an education program to convince the inhabitants that there was more money to be made from ecotourism than from hunting. Shortly afterward, he brought the first birders to the village. The project involved the entire community, so everyone could benefit. It was a resounding success. In the valley around the village there is hardly any hunting activity, and this example has since been followed by a number of other villages in the area.

We are welcomed with open arms. When we drive into the village, people are standing in front of their houses to wave at us. The children of our host family carry our backpacks to our rooms, the beds are made, and the women of the village prepare a feast. Every villager we meet greets us with a warm smile.

"You see what your visit brings about," Peter says proudly.

We go to bed immediately after dinner. We will be able to get a normal night's sleep for the first time in over a week, because the best birding spot is nearby, just outside the village. As soon as my head hits the pillow I fall asleep like a log, only to wake up nine hours later. I feel reborn.

———

A few years ago, our young guide used to walk around with a shotgun; now binoculars are hanging around his neck. He clearly takes pleasure in pointing out the birds in his backyard. If a few Dutchmen travel all the way from the Netherlands to his village to see these birds, they must be worth protecting.

He leads us to the territory of a Naga Wren-Babbler, an inconspicuous brown songbird that scurries through the undergrowth like a mouse and can be found only in these hills. He also points out a pair of Yellow-rumped Honeyguides on an overhanging branch above a steep rock wall with a huge wasp nest. The Latin name for this bird family is Indicatoridae, which means "pointers." Some species, such as the African Greater Honeyguide, are known for guiding honey badgers, as well as humans, to bee hives. When they break open the hive to extract the honey from the combs, the honeyguide feasts on the larvae and beeswax, their favorite food. It's a unique relationship between humans and birds.

———

When we return to the village, we are approached by a wrinkled and stooped man with a walking stick.

"This is the village elder," Peter says, shaking the man's hand and introducing him to me. "He walked down from his cottage on top of the hill especially to meet you."

We sit down, and one of the women pours us a cup of tea. With Peter as interpreter, the old man starts to tell his story. He explains how important this transformation from hunting to tourism has been to his village. The birds of his childhood were in danger of extinction, and the valley grew quieter and quieter. Now birdsong can be heard all around again. Respect for nature has returned to his community, he says. And that's because of Peter and the birders who travel to this place from far and wide.

After three wonderful weeks in India, Max and I say goodbye to our tour group and we set course for Thailand. I've been on the road for a month now, and my bird count stands at 792 species. Prior to my Big Year, I had calculated that, after India, it would have to stand at 800 species if I wanted to break Noah's record. So I am at least eight species behind schedule, and I will have to pull out all the stops.

Scheveningen Harbor, Scheveningen, Netherlands (1998)

In my last year of primary school, I became more and more fanatical about birds. Every day I cycled through the dunes, and my mother took me to the Wadden Islands whenever possible. I met hardly any other birders and certainly none of my age. I did not have the faintest idea what was rare and what wasn't. When I go back through my old bird logs, I frequently come across questionable rarities. Lack of knowledge, experience, and especially my rich imagination and wishful thinking often led to erroneous identifications. But hey, birding is like tennis: When you hold a racket in your hands for the first time, you don't immediately hit an ace. The ball will first end up in the net about a hundred times.

A stormy autumn day in September 1998 turned out to be decisive for my birding career. I was twelve and leafing through *Free Birds* magazine, which had just arrived in the mail. In the section "Birds of the Season," I read that seabirds were sometimes blown from the North Sea into the harbors of the Dutch coast during northwesterly storms. Through the window of the conservatory, I saw a dark pack of clouds speeding over our house from a northwesterly direction. Was this one of those days when gannets and Arctic Skuas could be seen in Scheveningen harbor?

I sprinted up the stairs to the study room.

"Mom! Mom! Can we go to the pier?"

My mother glanced outside. At that very moment, a gust of wind and rain passed by, bending the treetops. "But darling, why on earth do you want to go out in this weather?"

I showed her the seabirds section of the magazine, full of excitement. Convinced, finally, we drove off to the harbor. Together we walked onto the pier, my great-grandfather's old army binoculars dangling from my neck. At first, I saw only seagulls above the turbulent sea, and when we eventually reached the green lighthouse at the end of the pier, I was slightly disappointed. But there, sheltered by the lighthouse, a man in a green rain suit stood peering through a telescope on a tripod. He seemed to be concentrating deeply, and he looked up as I approached him.

"Hi, I'm Danny."

"What are you looking at?"

Danny pointed to his telescope. "Just take a look through this."

I squeezed my right eye shut and peered through the eyepiece with my left eye. At first I saw only whitecaps and a red buoy that sometimes half disappeared into the waves. But then it happened: From the right, a graceful, dark, gull-like bird with a pointed tail and a sharply contrasting white belly and neck flew into my field of vision.

"Is that a . . . ?"

"That's an Arctic Skua," Danny said.

I followed the skua breathlessly as he flew, apparently effortlessly, south through the troughs. I had never seen a seagull do that, flying so elegantly and gracefully.

31

"Do you see him making large turns while barely flapping his wings? It's called gliding," Danny said.

Wow, a gliding Arctic Skua, only a fifteen-minute bike ride away from my house in Scheveningen.

The rest of the afternoon I stood next to Danny on the Zuiderhaven pier. We saw gannets and Red-throated Loons, and he explained to me the difference between the Razorbills and Common Murres flying by. To top it all off, a Manx Shearwater skimmed over the waves. My mother came to pick me up that evening. My shoes were soaked with seawater, and I was shaking like a leaf in my summer jacket, which was far too thin. Nevertheless, I felt like the happiest child in the world.

Pak Thale, Thailand (2016)

Max and I arrive at the Pak Thale salt production area just after sunrise. My eyes are still dry from the uncomfortable night flight, and the clammy heat doesn't make things better. But there's no time to feel miserable; there's work to be done. And once again, it costs me surprisingly little effort. It seems as if, this year, I can draw on energy reserves that I normally do not have. Maybe it's the adrenaline rush of seeing new bird species, or the drive to set a world record that cannot be broken ever again.

I would have the chance to spot many new wader species here, but one of them rises head and shoulders above the rest: the Spoon-billed Sandpiper, sometimes affectionately referred to as "spoony" by birders.

Normally, sandpipers have a pointed beak, but the spoony has a kind of rounded triangular flap on either side of the beak tip, making its beak look like a teaspoon. With this "spoon" he hoes through shallow waters in search for food.

The most recent counts in the breeding and wintering areas indicate that there are fewer than 400 specimens left worldwide. The natural tidal areas in East and Southeast Asia are disappearing due to the development of cities along the coast. The large mudbanks and mudflats, used by the spoonies as stop-over sites during their yearly migration, have virtually disappeared, giving way to huge harbor complexes and industry. Moreover, the spoonies suffer great losses from the illegal hunting of migratory birds. Especially in China, wild birds are still caught and eaten on a large scale. In the Chinese province of Guangdong, so-called death zones have been found in a number of places: tidal areas where miles-long rows of mist nets have been set up, killing thousands of waders every single day.

The Spoon-billed Sandpiper breeds only in the vast tundra of Siberia's inhospitable (and inaccessible) Kamchatka and Chukótsk Peninsulas, making it extremely difficult for scientists to conduct research and implement conservation measures. The most important wintering areas are the vast mudflats off the coasts of Myanmar and Bangladesh, where these rarely occurring birds are virtually impossible to spot among the tens of thousands of waders.

Only a few spoonies overwinter at Pak Thale, but we have to be practical; these salt marshes are only a few hours' drive from Bangkok. Whether we will succeed remains to be seen. Only two birds have returned here from Siberia this year, as I read on the "Saving the Spoon-billed Sandpiper" website just before our departure.

As far as the eye can reach, we can see salt marshes with thousands of resting waders. It is almost impossible to identify them when their beaks are hidden between their shoulder feathers. We have only a few hours, because as soon as the low tide returns they will all take to their wings and head for the mudflats to look for food.

Our guide appears to have absolutely no knowledge of waders and has no idea where exactly to look. When I show him a picture of the Spoon-billed Sandpiper, he looks utterly puzzled. After walking and scouting like headless chickens for over an hour, a feeling of discouragement sets in. But then we bump into our savior: a British bird guide who accompanies a group of foreign birders.

"You look defeated. Can't seem to find a spoony?" he asks. "It must be your lucky day! We just found one. They have him in the telescope a bit further down."

We run to the telescope. In the middle of the field of vision, we indeed see a Spoon-billed Sandpiper, quietly having a brush up between a swarm of waders. Every now and then he removes his beak from his feathers and we can see the unmistakable spoon. It's a surreal feeling to watch such a rare bird, especially when you consider that it flew all the way here from Siberia. I run back to hug the Brit, who looks surprised, and thank him from the bottom of my heart.

There is still a glimmer of hope for this unique species. In a last-ditch effort to save it from extinction, BirdLife started a breeding program. In the breeding areas of Siberia, some of the eggs are carefully collected from the nests and incubated artificially. When the young spoonies have fledged, they are released in the vicinity of their nests and left to find the way to their wintering grounds on their own.

The first successes have been logged. Since 2013, a few spoonies that hatched in captivity—individually recognizable by the color-rings on their legs—have been spotted on the wintering grounds. Birds from the breeding program do indeed find their way to the tidal areas of Southeast Asia on their own. A few of these birds were found in the breeding areas in Northeast Siberia the following summer.

The Spoon-billed Sandpiper is still teetering on the brink of extinction, but I draw hope from these kinds of initiatives. Humans have destroyed a lot, but we also have the ability to learn from our mistakes and use our knowledge to turn things around.

Kaeng Krachan National Park, Phetchaburi, Thailand

We arrive at the camping site near the park headquarters by evening. After we have set up our tent, we decide to have something to eat in the only restaurant in the area: a wooden building with an open kitchen and a seating area filled with plastic chairs and lit by fluorescent tubes.

I'm aware that they like spicy food in Thailand, so when Max orders "chicken, extra extra spicy" for both of us, I can already feel it coming.

"You sure, sir?" asks the ranger, who is both waiter and chef, with a wide grin on his face, showing off his three remaining teeth.

"Sure, and make it extra spicy," Max replies.

Five minutes after we put in our order, we hear the man sneeze and cough as he stirs up our meal. The pepper fumes soon hit the eyes and respiratory tracts of everyone present.

The bowls that subsequently are placed in front of us look like plates of churning lava: a fiery red sauce containing some pieces of onion and chicken. With my first bite, it feels like I'm wolfing down a spoonful of red-hot coals. Sweat is beading on my forehead and, according to Max, my head has turned into a red balloon with white dots.

But we decide not to lose face. We will keep mum, no matter what.

"Not too spicy, sir?" the chef asks every so often.

Our shirts are soaked with sweat, and the few people in the restaurant are doubled over with laughter. My mouth is numb now. After a lot of pain and effort, we manage to fish the last pieces of onion and chicken out of the sauce and our ordeal is over.

"Not too spicy, sir?" he asks again with a wide grin.

Max looks him straight in the face. "It was delicious."

He then brings the plate to his mouth and drinks the churning lava in one go.

———•———

That night, stomach cramps keep us awake. The clammy heat makes it almost unbearable to stay in the tent. Moisture condenses on the inside of the canvas walls and drips down, forming small puddles in the corners of the tent. We have our earplugs in because our driver, who is sleeping in a hammock a few yards away, snores like a bear.

In the middle of the night, when I finally fall asleep, I dream that I hear a Bay Owl, but when I wake up and take out my earplugs, I can still hear it! I'm on edge: This is one of the most sought-after owls in the world. There are birders who have traveled to Southeast Asia a dozen times and have never managed to see one.

I quickly wake up Max. In order not to lose any time, we jump into our swimming trunks and walk in slippers to the edge of the forest, where the sound seems to come from. I play a recording of the owl through my speaker, in the hope it will get closer. But as befits this species, it calls back but refuses to move.

"This could be our only chance," I say defiantly to Max.

Armed with only a flashlight and binoculars, we walk into the pitch-dark rainforest in our slippers. Invisible thorns rip open our skin and we are stung by dozens of insects, but we are getting closer and closer to the sound. I almost manage to catch him in my light beam twice, but each time he manages to outsmart us. Our strategy doesn't work until our fifth attempt. The owl stares at us from a vine, at a distance of less than 10 yards. The light from my flashlight reflects a fiery red in his saucer-shaped eyes, which lie deep in his heart-shaped face.

Without any orientation point, it takes quite a bit of effort to find our way back in the dark forest. Fortunately there appears to be nothing wrong with Max's sense of direction—if it had been up to me, we would have been walking in the wrong direction—and eventually we reach the edge of the forest unscathed.

The next day the rangers in our camp think we are crazy. Rightly so, because that same evening a sun bear emerges from the forest, right next to the camp, and when we drive to the higher section of the park, a leopard crosses the road. Our operation of the night before was quite reckless. If I had known this, I would have stayed put. But then again, there probably wouldn't be a Bay Owl on my list right now. *The juice was worth the squeeze.*

The Volcano, Scheveningen, Netherlands (1998)

In school, I was addicted to bird migration monitoring—the counting, observing, and recording of birds passing by during their migration. Almost every day before or after and often even during school, I could be found on the Zuiderhavenhoofd or on the highest dune top in the Westduinpark, better known among bird-watchers as The Volcano.

I soon discovered that Danny was not the only birder in The Hague. There were a few dozen like-minded people, and all of them preferred to spend their spare time standing on a windswept dune top or on the pier of Scheveningen.

I really looked up to Vincent and Rinse, two men in their early twenties who smoked cigarettes, listened to grunge music, and talked about women all the time; however, they were also excellent birders. They thought I was a pain in the ass. Nearly every day, that blond, annoying, spoiled, twelve-year-old little fellow talked their ears off nonstop and exasperated them with his silly questions.

"Which is rarer? A Snow Bunting or a Lapland Bunting?"

"What do you like better? A Common Greenshank or a Spotted Redshank?"

"What's bigger? A buzzard or a Honey Buzzard?"

Vincent then grabbed my backpack containing my Thermos and packed lunch and threw it as far as he could into the sea buckthorns at the bottom of The Volcano.

"Here, go and see which hurts more, a nettle or a sea buckthorn," he said with a big grin on his face.

When, as a child, you pursue a hobby so intensively, you soak up information like a sponge. I learned how to recognize different species of passing songbirds by their calls, how to correctly estimate the group size of a flock of thousands of starlings, and how to determine soaring birds of prey by the position of their wings from a great distance. And I put up with Vincent's and Rinse's teasing; they would go on to become two of my best birder friends.

Kinabatangan River, Sabah, Malaysia (2016)

Exactly ten years before my Big Year, Borneo was the first stop on my first world trip. Even back then, a large part of the oldest rainforest in the world had been cleared to make way for palm oil plantations. I pray that the damage has remained limited after all these years, but I fear the worst. More than 50 percent of the products in our supermarkets contain palm oil, and this has been the case for decades. Peanut butter, cookies, biofuel, shower gel, chips, even baby food . . . you name it, all contain palm oil. So it's not only the obvious culprits like livestock farming, fishing, illegal logging, poaching, pollution, and our use of fossil fuels that are killing nature; almost all facets of our modern Western life contribute to the destruction of nature.

———————

At the village of Sukau, which is situated along the Kinabatangan River, Max and I board a boat to the Kinabatangan Jungle Camp. We are joined by Sander and Jelmer, two birders I've known since I was twelve years old. I met them during my first autumn camps on Schiermonnikoog, one of the Frisian Wadden Islands. Sander pointed out my first Yellow-browed Warbler, a tiny bird from Siberia, which somewhat resembles a firecrest due to its striking eyebrow stripe, and Jelmer joined me when I traveled through Peru for five weeks during my world trip in 2006.

When we sail to the lodge, I notice that the forest along the river has remained fairly intact, but where ten years ago large numbers of hornbills were flying around, we now only see one. During our first boat safari, the owner of the lodge tells us that the seemingly pristine rainforest along the river is just a facade. Much of the forest in the hinterland has been cleared, making way for palm oil plantations. In some places you can even see through the forest, as if it were a transparent shower curtain. It's relatively easy to spot rare animals like Borneo pygmy elephants and orangutans here, but that's because these animals have nowhere else to go. They are forced to live in a narrow strip of rainforest, only a few miles wide at best, as there is hardly any food to be found in the adjacent palm oil plantations.

The hornbills have another problem still. They use the oldest and tallest trees to breed, and it is mainly these trees that are felled illegally. The situation is most critical for the Helmeted Hornbill, the largest of them all. In Kalimantan—the Indonesian part of Borneo—these impressive birds are poached and sold to China, where they make figurines with their massive red-and-yellow-colored upper mandibles. This "horn" is called "yellow ivory" on the black market.

———————

In the late 1990s, the Kinabatangan Orangutan Conservation Programme was created with the aim of restoring the orangutan habitat and creating corridors between the remaining stretches of forest along the river. These natural "bridges" are important for the genetic exchange between different animal populations. In the rainforests of Borneo, orangutans are highly dependent on fruit trees, so it is very important that they are able to freely migrate between forest patches, otherwise they would be forced to cross palm oil plantations or open logging areas, where they are vulnerable to accidents and poaching.

While Sander, Max, and I have an afternoon break and are relaxing at the lodge, Jelmer suddenly comes running toward us. "I've seen an orangutan!"

We run after him, and a moment later the four of us are watching the long-haired, reddish-brown ape feasting on the ripe fruit at the top of a fig tree. With a twisting motion, he pulls the figs off the branch and puts them in his mouth, one by one, just like a human would do. It is a young male: He does not have the characteristic round cheek plates of an adult animal yet. When I study his facial expression through my binoculars, it's easy to fathom that his DNA differs from mine only by 3 percent; the look in his eyes is so incredibly human.

Slightly less known, but no less unique, is the proboscis monkey. The locals call this monkey "orang Belanda"—Dutch man. Its swollen nose, round belly, and orange fur strongly reminded the local residents of the first Dutch sailors who set foot ashore in the seventeenth century. We see this bizarre-looking monkey in the downstream mangrove forests, in family groups of ten to forty animals. They feed almost exclusively on the young, green leaves of mangrove trees. This food is so hard to digest that they spend most of the day chewing their cud, which, in combination with their droopy nose; tiny, round, beady eyes; and their ever-stiff, bright red willies, gives them a somewhat comical appearance.

As we sail down the river, it is slowly getting dark. We are looking for the Large Frogmouth, a giant nocturnal bird with a distinctive broad beak, that looks like something between an owl and a nightjar. Just before dark, the sound of cicadas swells to a deafening noise, then tree frogs and crickets take over. Our guide turns off the boat engine, and the waiting begins. Suddenly, a bizarre hollow scream echoes across the river.

"That's the frogmouth," whispers the guide.

He puts his hands around his mouth and to our surprise produces a perfect imitation of the sound we just heard. Seconds later, a dark shadow flies from the other side of the river, straight over our heads and into the forest.

"Wait here." He jumps ashore and disappears among the riparian vegetation. Every so often, the frogmouth calls and the guide answers. We see a beam of light moving searchingly through the treetops. After a few attempts, the beam remains focused on one point. A soft whistle confirms that he has found the bird, so we jump out of the boat and walk over. The frogmouth is less than 5 yards away, clearly visible in the beam of the flashlight. We can admire its auburn, camouflaged plumage and the unique, hair-shaped "eyebrows" above its eyes, as it occasionally bites at the moths that flock to the beam of the flashlight.

Danum Valley, Sabah, Malaysia

The Danum Valley is an immense tropical rainforest, the largest in Southeast Asia. Every morning before dawn, we head up the network of forest trails around the research station. It is important that we are in the best spots when the first light breaks through, as, unlike the rainforests in South America and Africa, in Southeast Asia's rainforests, all bird activity completely stops after about nine o'clock in the morning. At that time of day, the sun starts to burn mercilessly, and the omnipresent noise of cicadas drowns out all bird sounds. It is wise to keep a low profile during the hottest part of the day. The rewards in the forest are then so low that it is a waste of energy to search. However, I'm attempting to break a world record and hence, we break that rule of thumb every single day. All day long we trudge along the sweltering forest paths. Sometimes we don't see anything at all for hours on end, but from time to time our efforts pay off with a new bird species. If I want to break Noah's record, I have to go all the way, no matter where we are. There will be time enough to rest next year.

By evening, we decide to try our luck along the access road. The temperature has dropped considerably, becoming somewhat bearable. Unsuspectingly, we pass a bend in the road when suddenly, at a distance of less than 20 yards, we see a herd of Bornean pygmy elephants. I freeze, for I know the reputation of the Asian elephant; a few years ago, a Dutch birder was trampled to death during a trip in Northeast India.

They may be called pygmy elephants, but this subspecies is barely smaller than a normal Asian elephant.

"Do not move and quietly walk backwards," Max whispers. He has worked as a ranger in Namibia and has experience dealing with these kinds of situations. His warning falls on deaf ears, as I'm already gone, running in the other direction in a blind panic.

Fortunately, there is a lookout tower on a hill next to the road. The elephants are standing between us and the research station; walking past the herd would have been complete suicide. From this safe point, 32 feet above the ground, we try to call the research station. But we don't have enough coverage, and it is impossible to get a connection.

While we are in the tower, Max points down. "This may get ugly."

A few yards from the herd, two women are taking *selfies* with the elephants. I can hardly believe it; one wrong move and they're dead. We wave at them and signal with our laser pointers. They do see us but ignore our warnings and decide to remain where they are, oblivious to any danger. In the end, it's too dark to take pictures and they walk away. We breathe a sigh of relief. We have to wait another two hours before a 4×4 finally drives by and gives us a lift. Apparently the ladies didn't even bother to warn anyone in the research station.

When we walk into the dining room a little later, they are standing at the buffet.

"Do you have any idea how dangerous that was?" I ask.

One of them looks at me sheepishly.

"Oh my god, you guys were seriously freaking out," she says in a heavy California accent. "We are very good with elephants. We always get on fine with them in the San Diego Zoo at home. If you're not scared, these animals won't hurt you."

I open my mouth to start a discussion, but then realize that this would be the same as talking to a wall.

Amsterdam Dunes, Netherlands (2002)

Most birders in The Hague had a Leica or a Swarovski around their neck. I walked around with a cheap pair of binoculars that was so out of focus it hurt my eyes. On The Volcano, I cast envious glances at the binoculars of the other birders. With that prehistoric thing of mine, I felt like a clown, and that's how I was treated by Vincent and Rinse. That had to change, so I took on a newspaper route.

On a wintery day in January, my day had finally arrived: I cycled to Combi Focus with 1,800 guilders in my pocket. A little later I could call myself the proud owner of brand-new binoculars. Everything I saw through these binoculars seemed to turn to gold: the plane contrail high in the air above our garden, the Dutch flag on the Kurhaus, and the jackdaw on the neighbors' chimney.

"Shall we go to the Amsterdamse Waterleidingduinen this afternoon?" I asked my parents. There were reports of a dipper being spotted there, and I had never seen one. And wouldn't it be great if I could lay my eyes on a new species through my new binoculars?

The dipper appeared to have other plans, but I did not care. Like a rapper with a 24-carat gold chain, I paraded through the dunes with my brand-new accessory. Finally, I could consider myself a real pro.

On the way back to the parking lot, I saw something scurrying through the undergrowth. I aimed my binoculars and saw a small songbird with striking yellow stripes on its head and two small stripes on its wings, moving through a bush along with a Goldcrest. I could hardly believe my eyes: I was watching a Pallas's Leaf Warbler! It was a Siberian vagrant and one of the very first winter sightings in the Netherlands.

The warbler lingered in the dune area for another week, and many other birders enjoyed my discovery. From that moment on, I credited my binoculars with magical powers, a bit like Charlie's Wonder Slippers.

A few days later, when I was standing next to Vincent on the pier, my new binoculars caught his eye.

"Even if a monkey wears a gold ring, it is and remains an ugly thing," he said, roaring with laughter.

Davao, Mindanao, Philippines (2016)

The Philippine Eagle is arguably the ultimate bird. It is the most powerful eagle in the world, with a wingspan of more than 7 feet. Its talons and nails are the size of a grizzly bear's paws. It has a gigantic light blue beak, piercing light irises, and a fluttering blond crest. In Dutch, it is called a Monkey Eagle because it hunts monkeys—although the colugo, or flying lemur, is its favorite prey. It is the national bird of the Philippines and according to the latest BirdLife estimate, only 180 to 500 of these birds are still remaining in the wild. The first nest was discovered only in 1963.

They live almost exclusively on Mindanao and Luzon, the two largest islands of the Philippine archipelago. A pair of these eagles covers a territory of more than 80 square miles, more than half of which must be primary rainforest. Due to the large-scale deforestation in the Philippines, such untouched pieces of nature are becoming increasingly rare. In addition, a female doesn't become sexually mature until after five years and a male after seven years, and a pair raises only one young every two years at most.

BirdLife International and the Philippine Eagle Foundation are working closely together to save the Phillipine Eagle from extinction. Local communities are involved in sustainable ecotourism, nests are being protected from poachers in every way possible, and education programs should help Filipino children understand the value of these birds. But it remains a race against time.

Shortly after our departure from the Netherlands, it became clear that our chances to actually see a Phillipine Eagle were very slim. According to our local contact, Pete, two of the most accessible locations were no longer safe due to rebel activities, and there were no known active nests at the time. Max and I resigned ourselves to the fact that we probably weren't going to see this "species of all species."

Two weeks before our arrival in Mindanao, our luck changed: I received a message from Pete that a juvenile Phillipine Eagle had returned to its nest after more than six months (juvenile Phillipine Eagles can spend more than six months lingering in the vicinity of their nest). Through the Philippine Eagle Foundation, he had managed to arrange for Max and me to visit this location with special permits.

Our adventure starts in the back of a pickup truck. We drive out of town and, after a few more hours of jerking up and down, climbing uphill on a mostly unpaved road, we arrive at a small village in the hills. This location provides us with an overwhelmingly beautiful view. Far below us is the Pacific Ocean. The coral reefs off the coast stand out slightly in the cerulean water. A snow-white beach of coral sand forms the border with a green blanket of palm trees, between which we can perceive the thatched roofs of houses.

Our documents are inspected in an office, and we are asked to make a financial contribution. Some of it goes to the villagers, encouraging them to protect the nest, and some goes directly to the Philippine Eagle Foundation for its breeding program.

Accompanied by a young local guide, we walk up a narrow path past fields and patches of forest. There is no indication that we are approaching a Phillipine Eagle's nest. At a lookout point on the edge of a wooded valley, the guide pulls my arm and points to a gigantic tree 65 feet below us. In the crown of the tree I see a huge aerie. My heart rate goes through the roof. The chances of us seeing a Phillipine Eagle any minute now are greater than I ever thought possible.

The adrenaline rush slowly ebbs away. Other than the huge nest, we have no evidence whether the eagle is indeed anywhere nearby. The only sign of life comes from a group of macaques that are quietly grooming each other in an old tree across the valley. If their nemesis had been around, I'm sure they wouldn't have been so blissfully insouciant.

Hours pass by during which we see nothing. The sun is blazing in full force, and the macaques have made themselves comfortable underneath the tree's shady canopy. One is stretched out on a branch, letting its four arms dangle freely, without a care in the world.

"You have to be patient," the guide says. "He usually doesn't come until late in the morning, when it's a bit warmer."

Just then, panic breaks out among the macaque family. They scream and jump back and forth. The crown of the tree shakes in all directions. From the right, a gigantic bird of prey sails into the valley.

"*Phillipine Eagle!*"

We don't have to doubt this for a second. The bird has an enormous wingspan, but with its long head and tail and a heavy chest, its physique bears a greater resemblance to a hawk than a Bald Eagle.

The eagle lands in a tree across the valley. Despite the distance, we can have a good look at it with my telescope zoomed in sixty times. The entire underside of the bird is bright white, and it has a flowing blond tuft. His claws are huge, making it look like he's wearing boxing gloves. The equally huge beak is light gray, so it must be a young bird.

Our euphoria soon turns into concern: He clearly has an injury to his right eye, and when he opens his wings, there appears to be a wound to his left wing. Could he have been shot?

The persecution of birds of prey is a major problem in the Philippines. Phillipine Eagles and other birds of prey are blamed when chickens or piglets disappear from yards. Almost every year, an eagle is found killed by gunshot hail—a huge impact for a species that raises young only once every two years and of which there are so few left.

As we watch the bird through the telescope, it begins to squeal, "*Kiuu, Kiuu, Kiuu!*" Almost immediately an answer is heard from the right, and to our great surprise, a second eagle with a light blue beak appears: an adult specimen.

"This is the mother," the guide says. "The father was shot dead a few months ago."

His words hit us like a bomb. What madman would dare to point a gun at such an awesome animal and pull the trigger?

Bislig, Mindanao, Philippines

When Max and I meet our guide, Zardo, at the Paper Country Inn, the only small hotel in the town of Bislig, it is three in the morning and pitch dark. It will take an hours-long journey to reach the best birding spot and we want to get there before sunrise. We leave in the back of a jeepney, the most common mode of transport in the Philippines. It's a kind of elongated tin 4×4 that constantly emits dark clouds of soot—particulate filters don't seem to exist here. The jeepney has little or no suspension, causing us to bounce against the ceiling at every bump or hole in the road surface.

We head for the lands of Paper Industries Corporation of the Philippines (PICOP), one of Mindanao's last remaining stretches of lowland rainforest. This is the best-known birding destination in the Philippines, but even this legendary place is suffering from deforestation.

In the early 1950s, a huge logging concession of nearly 500,000 acres was granted by the Philippine government to the company Bislig Industries, which was renamed PICOP ten years later. The company applied selective logging: Only the largest and oldest trees were felled, and the rest of the forest was left intact as much as possible. Because PICOP had a monopoly, deforestation was relatively limited here, while almost all of Mindanao's other lowland rainforest was gone in less than half a century.

PICOP went bankrupt at the end of the 1980s and from that moment on, the illegal logging industry was given free rein. Opportunists moved into the area, chopping, sawing, and felling while crisscrossing the jungle, leaving a trail of destruction in their wake. Today, PICOP's lowland rainforest is a mere shadow of what it used to be. The unique bird and animal life here depends on the last remaining clusters of forest amid smoking clearings—after the largest trees have been felled, the remaining forest is burned to charcoal.

In the past, the rainforest virtually bordered the city, now we have to drive a few hours to reach the remaining forest patches.

We arrive at our destination just before sunrise. There is a lot of bird activity because the weather is good—cloudy with a short but heavy shower now and then, a weather pattern that keeps birds active throughout the day. We see one rarity after another, such as the Azure-breasted Pitta and the Wattled Broadbill.

At any other moment I would have been over the moon after such a great score, but the devastation of nature visible all around us leaves a bitter aftertaste. All day long the noise of chainsaws can be heard in the background. The nerve-racking sound is harrowing, and every tree we hear being felled makes me feel like I'm being punched in the stomach.

After two long days in the field, we arrive at the Bislig airport around midnight. Our flight to Manila doesn't leave until 6 a.m. The plan was to save money by avoiding a hotel and getting a few hours of sleep in a quiet spot in the departure hall. But it turns out that nothing is open until 4:30 in the morning. We have no choice but to inflate our air mattresses and find a place to sleep in the street near the entrance. We take turns trying to get some sleep for an hour, while the other keeps an eye on things. Meanwhile, we are attacked by hordes of bloodthirsty mosquitoes.

"It can't get much worse than this," Max says, giving me an exhausted look from his air mattress.

When he wakes me up after my first hour of sleep, my lips are severely swollen and itching like mad. I had covered myself with DEET from head to toe, but I forgot to put something on my lips. Another lesson learned.

Now it's my turn to stay awake. I use the time to write my blog and to review my sightings on my phone. When I look at the long list of bird species from the past few days, I conclude that all our hardship has been 100 percent worth it. I would do it again. With all the wild-life devastation in the Philippines, it is conceivable that some of these species will be extinct in the not-too-distant future. I can't even bring myself to think I could be proven right.

Scheveningen, Netherlands (1999)

During the early years of high school I was a bit of a loner. Birding was simply not cool in the eyes of teenagers. During breaks and between classes, my classmates gathered around the foosball table in the cafeteria. Since I knew it would never be my turn anyway, I just walked around the neighborhood on my own until the redeeming sound of the school bell rang and class started again. When I had the first two hours off school and cycled back home after counting migratory birds on the pier or The Volcano, I always made sure that my binoculars were safely tucked away under my jacket. Imagine if a girl in my class saw me with that thing. Only the tripod on the back of my bike could reveal what I had been doing.

Like many of my peers, I played field hockey, where I shone as the goalkeeper who was afraid of the ball. On Wednesday afternoons we had training, and on Saturdays we played competition. The prospect of drag pushes and C-cut pushes being fired at me for more than an hour made me reluctantly head off to the hockey club every Saturday with a healthy dose of resentment. What made it even worse is that I would miss an entire afternoon of birding. During the match, I kept an eye on the sky. It could just so happen that a Black Stork or Purple Heron might suddenly fly over on a Saturday afternoon in September.

"Arjan, watch out!"

I woke up from my daydreams, and my imaginary bird migration station turned back into a hockey goal. Just then, I saw a ball pass by me painfully slowly and roll into the goal.

"Idiot! This is where it happens, on the field, not in the air," our coach shouted, while the referee noted the 0–1 disadvantage in his scoring book.

Java, Indonesia (2016)

"We'll start walking at four," Khaleb, our guide, states before we go to bed.

The rain is pouring nonstop and makes a deafening noise on the corrugated metal roof of our small hotel. I can't fall asleep. Partly because of the noise and partly because of the prospect that we will soon have to walk up a mountain in the dark, through a soaking wet rainforest. What a contrast to this morning, when we were still walking around in our swimming trunks on a paradise beach, watching a Palawan Peacock-Pheasant.

I had added Gunung Gede Pangrango National Park to my itinerary at the last minute. Not without risk, because the beginning of March is the middle of the rainy season here and, furthermore, Gunung Gede Pangrango is known to be a notorious wet spot. Although birding in a tropical mountain forest is not easy in itself, it is almost impossible when rain comes pouring down from the heavens. To make things more difficult for ourselves, we will also set up camp for two nights in the middle of the park. It's no secret that pitching a tent when everything is soaking wet is disastrous for morale.

~ — ~

Due to the high humidity and the marching pace of Khaleb, my poncho is soon just as wet inside as it is outside. The morning light is slow to break through, and it turns out to be difficult to watch birds through our dripping binoculars.

When we arrive in the cloud forest halfway through the afternoon, the rain has turned into fog and drizzle. The environment is magnificent. We see forested slopes all around us, and we pass by sparkling waterfalls that plunge down between a sea of greenery. Mossy branches of ancient trees provide support for bromeliads and brightly colored orchids. Our camping spot is a stone's throw from a hot spring, from which steam swirls up and drifts away through the treetops.

Toward the end of the afternoon, it dries up for a while, revealing a pair of Javan Trogons, plump birds with long tails, blue-green backs and short, red beaks. Like so many birds in this park, this one occurs only in Java. They sit perfectly still on a branch, with their backs to us, and only when they turn around can we see their bright yellow bellies with a moss-green breast band.

In the evening, we eat our instant noodles by the light of a flashlight. A rustle in the bushes is caused by a Javan ferret-badger, a black-and-white mammal that somewhat holds the middle between a polecat and a badger. After a while, his greed beats his shyness and he throws himself onto our plates of leftovers.

We find ourselves in the middle of nature, far from civilization, with only the sound of ticking rain and the soft call of the endemic Javan Scops Owl to fall asleep to. It doesn't get any better than this; no five-star hotel in the world can compete.

It is still pitch dark when the alarm goes off. When I draw down the zipper and stick my head out of the tent, I'm surprised to see a crystal-clear sky with thousands of stars. It has stopped raining. We quickly clear up the tent because we don't want to miss a minute of the dry weather. As we go, Khaleb notices the soft, ethereal whistle of a Javan Cochoa, one of the toughest species to spot on this mountain. We immediately stop packing and follow him. Suddenly we see the bird. In the semidarkness I shine my flashlight so we can see its pale blue crown and wings. A perfect start to the day!

Keeping cage birds is immensely popular in Southeast Asia. The more beautiful the appearance and song of a species, the more popular it is as a cage bird. Many of these birds are not captive-bred but are caught in the wild. In Java alone, tens of thousands of illegally caught birds are traded each year, including a large number of critically endangered species. I heard from a birder friend who travels a lot through Indonesia that some rainforests on Sumatra and Java are now completely silent. The birds that once sang there are now languishing in cages. This is especially disastrous for the Rufous-fronted Laughingthrush because it naturally has a small range, limited to a few isolated tropical mountain forests in West Java. Its rarity makes it a favorite target for criminal organizations. BirdLife started a breeding program to save this laughingthrush from extinction, but just before I left on my trip, they had an armed robbery and three birds were stolen.

Even hefty fines and prison terms don't stop criminals from dealing in endangered species; it's simply too lucrative a business. It is, therefore, important to tackle the problem at the source. Education programs make children aware that birds belong in the wild and not in a cage. And that's exactly what BirdLife does, because when the demand disappears, the supply will eventually disappear, as well.

While we walk on a narrow, steep path through the forest, we hear a group of laughingthrushes calling far below us. We run down the muddy path like madmen. Hundreds of feet lower, we come to a halt, gasping for breath. We suddenly hear the sound, now much closer. We peer through the foliage with our binoculars. There is a brief movement in a bush diagonally above us on the slope. I nudge Max and we focus. A minute that seems like an eternity goes by, then a Rufous-fronted Laughingthrush leaps out of the undergrowth. It lands on a thick, mossy branch and sits fully exposed for a moment, giving us a good view of its distinctive reddish-brown forehead and yellowish eyes, and then, as suddenly as it appeared, it dives back into the dense vegetation.

Palu, Indonesia

I eagerly look forward to a shower, because I've run out of clean laundry and have been wearing the same clothes for almost a week. By now I smell like both a wet dog and a carton of sour milk. My cap is constantly soaked with sweat, making my hair look like spaghetti, and my socks seem to have grown into my feet. I didn't dare take my shoes off during the plane trip for fear of giving our fellow passengers airsickness.

It is around midnight when we arrive in Palu, a town in the north of Celebes. Our guide, Nurlin, picks us up and immediately makes it clear that we better put the idea of a shower and a bed out of our heads.

"We will camp deep in the jungle tonight so we can start looking for the Maleo in the early hours tomorrow morning."

After two hours in a van, we arrive at the edge of the jungle, where three off-road motorcycles are waiting for us. Before I know it, I'm on the back of a sixteen-year-old boy's motorcycle, racing at 40 miles per hour on a forest track barely 8 inches wide. The light from the head-lamp reveals that we are driving through a cocoa plantation. I have to keep my head down to avoid getting hit by branches and low-hanging fruits the size of rugby balls. Max falls off twice and ends up in the mud with all his luggage.

After an hour of agony on the back of the bikes, we arrive at a clearing on the bank of a fast-flowing river.

"We have to wade across with our luggage," Nurlin says, pointing to the churning water. "And from there, it's only half an hour's walk to our camp."

We take off our shoes and socks, pull up our pants, and wade to the other side of the river with our stuff on top of our heads. I can hardly stand on the smooth boulders in the fast-flowing water, and a few times I almost fall, luggage and all, into the river.

We continue on with wet feet. Soon the adrenaline disappears and fatigue sets in. I almost collapse when we finally see the soft glow of oil lamps in the distance: We have reached the encampment. And who are waiting for us there? Michiel and John!

Medemblik, Netherlands (2012)

During my student days, I saw the name Michiel van den Bergh for the first time when it appeared, exactly one place above me, on the Club 4500 ranking, with 4,321 species observed. He was only a year older than me and planned to spend three months in the interior of Papua New Guinea for his master's degree research.

Up to that point, I didn't know anyone my age who had been there. And moreover, he would travel there on his own, to do research. I became intrigued by this Michiel.

A few months later, he had climbed in the ranking with hundreds of other species observed.

"Highlight of three months of research was a displaying male Black Sicklebill," his latest update read. I was eating my heart out: The Black Sicklebill can be found only in the cloud forests of New Guinea, and it was one of my most-wanted species.

It took two more years before Michiel and I first met. It was in October 2012, on a windy dike near Medemblik. There were dozens of birders, who, just like me, were waiting for the fourth Pallid Swift ever to be recorded in the Netherlands, which had recently been discovered and photographed at this location. Due to the long wait, many birders lost their concentration and started to tell tall stories about long journeys, legendary discoveries, and rarities from a distant past.

Suddenly someone next to me roared loudly: "*Richard's Pipit!*"

The chatter died down and a few seconds later we could hear the distinctive frothy "*pschee!*" from a Richard's Pipit flying by.

"Wow, you spotted that sharply."

He shook my hand and introduced himself as Michiel.

———

We became good friends. We could be regularly found on the pier of IJmuiden, and we went together on birding weekends to Texel and Schiermonnikoog islands. We made long days in the field and spent our evenings in the pub, with sunburned faces and a glass of beer in our hands.

Two months before the start of my Big Year, he called me.

"I have big news. I'm going to make a documentary about your journey. I don't know how yet, but it would be a shame not to capture this adventure on film."

I didn't take his plans too seriously at first, but two weeks before my departure he called me again. "We're lucky. I managed to find a cameraman," he told me.

That same evening, Michiel, cameraman John, and I raised a glass to the documentary *Arjan's Big Year*.

Celebes, Indonesia (2016)

Despite being overtired, I can't fall asleep. We lie on the bare ground under a plastic sheet, and every time I breathe I can smell myself, a pungent, musty ammonia smell. Everything I'm wearing is damp, and white circles of salt have formed around the armpits on my T-shirt. Our mosquito net also serves as a blanket. The thin mesh lies directly on my skin and, hence, offers hardly any protection against mosquitoes, which sting through the fine mesh with ease. Fortunately, this hellish night is short-lived. We didn't go to bed until four o'clock and the alarm clock is set at a quarter past five. Michiel and John have been filming here for two nights and have barely had any sleep.

We traveled to this remote place to see the Maleo, and we hope to capture it on film. This black-and-white fowl-like bird, with its giant blue knob atop its crown, reproduces in a unique way. It does not incubate its eggs itself but makes use of the heat of the earth. A Maleo female lays eight to twelve eggs in a hole, which she digs deep in the sand. The location is important because it has to offer just enough geothermal heat. Too little and the eggs won't hatch, too much and they will become hard-boiled.

Beaches of black volcanic sand and sandy bottoms near underground hot springs are its favorites. A nesting site can be used for decades by dozens of Maleos coming from miles around. When the eggs hatch two to three months later, the young birds dig their way out and disappear into the forest. As soon as it hatches, a Maleo chick is completely independent.

We quickly eat a banana while we dress ourselves in the light of our headlamps. In the background, we can hear the monotonous "*Hoo-hoo-hoo-hoo-hoo*" of an Ochre-bellied Boobook, an orange-brown owl that, like the Maleo, occurs only on Celebes. I look at Michiel and John. They have huge bags under their eyes, and their skin is gray with fatigue. The lack of sleep has clearly affected them.

When the first light breaks through, we walk up to the nesting site. It's important that we keep quiet, because Maleos are incredibly shy and disappear at the slightest disturbance. We walk in silence, sneaking behind Nurlin down the trail. Max and I are in front, while Michiel and John, who are filming everything, trail behind. Next to a huge cage on the edge of a clearing is a sign with a BirdLife International logo and the text MALEO BREEDING PROGRAM. When we get closer, we see a number of adult Maleos and dozens of chicks, at most a few days old, walking around. As soon as they see us they panic and dash away to the farthest corner of the cage. Unfortunately, they do not count for my Big Year, as these birds live in captivity.

BirdLife has established a breeding program to save this species from extinction. Cages have been placed next to a number of known nesting sites. The wild-laid eggs are carefully dug up and buried inside the cage, where they are safe from hungry monitor lizards and wild pigs. Later, the chicks will be released into the wild.

We continue our sneaking journey. There is a strong smell of rotten eggs in the air, the smell of sulfur. Every now and then I see clouds of sulfur vapor curl up from the ground. The ground beneath our feet is volcanic, and the heat from an underground hot spring creates the perfect conditions for a Maleo nursery.

Nurlin motions us to sit down, and we wait in absolute silence. Hours have gone by, and we still haven't caught a glimpse of a wild Maleo. The silence is broken by a Knobbed Hornbill, an impressive yellow-black hornbill with a huge round red comb on its beak, which settles down above our heads in an old fig tree.

I'm starting to get quite worried now. The chances of seeing this extremely shy bird in the middle of the day are pretty slim.

Around noon, we have to give up, and we feel somewhat defeated when we drag ourselves back to our encampment. After we have packed our things, we decide to make one more attempt.

As we walk back to the clearing and pass by the cage, I suddenly hear Max shouting. Judging by the excitement in his voice, it can mean only one thing: He's got one in view! In a few seconds I'm at his side. After a few quick directions, I spot them: two adult Maleos, right next to the cage. The blue knob on the top of their heads and the yellow skin around their eyes give the birds a prehistoric look. They have undoubtedly come to see their conspecifics in the cage.

We are ecstatic! These wild birds look exactly the same as the birds in the cage, but the adrenaline rush that comes with the sighting makes them many times more beautiful. This might be hard to understand if you're not a birder. You can compare it to a live concert by your favorite artist: That probably triggers a lot more in you than seeing something live on television.

Afterward, Max and I decide to symbolically adopt two chicks. After paying a few euros, which go straight to the Maleo breeding program, we are allowed to release the chicks into the wild. Will they ever return to this place, like their mother, to bury their eggs in the volcanic soil?

After a long journey back to civilization, we continue to Lore Lindu National Park. This park contains by far the most endemic species on Celebes. It is the site of a famous path, the Anaso Track, at the top of which we hope to find a rare, thrush-like bird, the Geomalia, which occurs only on the forest floor of the cloud forests of Celebes.

We arrive at our hotel in the evening. Finally we can hand in our laundry. The Indonesian lady smiles kindly but holds our dirty clothes with her arms stretched forward, as far away as she possibly can. She pulls a grimace and makes a gesture as if she is pinching her nose.

Never ever have I enjoyed a lukewarm shower as much as today. A bit of shampoo turns my spaghetti hair back to normal, and the cool water relieves the mosquito and aphid bites on my body.

We feel reenergized as we walk up to the restaurant, when we suddenly hear the call of a Sulawesi Masked Owl. Shortly afterward, a large white Barn Owl flies by and lands at the top of a tree canopy. I quickly grab my flashlight and bingo! From a bare branch, the dark-eyed owl stares at us stoically. There's something sinister about him, like a messenger from the underworld. It's not for nothing that Barn Owls appear over and over again in all kinds of mythologies and folklore; often they are associated with death.

The Masked Owl lives up to its reputation as a bearer of bad news: The next morning, Nurlin announces that the Anaso Track is closed. There is fighting between the Indonesian army and Islamic State–affiliated militants, right at the location where the Geomalia and a number of other unique species occur. The leader of this Islamic movement is the mastermind behind the deadly bombings in Jakarta, which took place just before my departure; seven people were killed and dozens of people were injured. Radical Islam has been gaining traction in Indonesia in recent years, and tensions between the moderate Muslim population and the more radical Wahhabists are rising, resulting in these terrible attacks.

Twice we pass by a path that is marked off with a red-and-white ribbon. The mossy sign reads ANASO TRACK. Behind the ribbon, the path winds steeply up. I close my eyes for a moment and can't help but think of the Geomalia, somewhere up there, wading through the undergrowth of the cloud forest, while in the background the firefight between the rebels and the army erupts in full force.

"I have to think very carefully about whether I am still willing or even able to continue with this project," John states after we have arrived at the airport in Makassar.

He is about to collapse. For John, who has to carry around a heavy camera and tripod everywhere, the adventure turned out to be a real test of endurance. Every night when we went to bed, he had to stay up for two hours to charge his batteries and transfer images to a hard drive. And every morning he had to get up an hour earlier to get all his equipment ready, while our nights were already too short.

He and Michiel are going back to the Netherlands. If John decides to continue with the project, they will visit me again in June, in Ghana.

"You did a good job, buddy," I say to John, wrapping my arms around him. "You will see that it will be a hundred times easier in Ghana."

I'm not telling him that West Africa will be in the middle of the rainy season, which could make filming a documentary a hellish experience. First, John clearly needs some time to let the whole Celebes adventure sink in.

Michiel gives me an encouraging slap on the shoulder. "Go for it, dude."

Max and I walk to our gate for the long journey to Papua New Guinea.

Scheveningen, Netherlands (2002)

As a teenager, I sometimes found it quite hard that my love for birds was ridiculed. Why was I being bullied for something I enjoyed so much? During free periods, my classmates grouped together to smoke cigarettes, listen to dance music, and talk about going out and girls. I hated dancing, didn't go out, and wasn't yet interested in the opposite sex. At least that's what I told myself. I sometimes tried to join such a group, but after a few minutes I usually couldn't take it anymore. Then I would cycle home to pick up my binoculars and go into the dunes.

My school performance also suffered from my hobby. I'd rather be on The Volcano than doing homework. When the wind was favorable for bird migration, I often played hooky from school for half a day, and during class I'd rather stare out the window than listen to the teacher. In the fourth grade, I received six unsatisfactory marks and had to repeat a year. My parents were disappointed, but I was secretly happy. I wouldn't have to work so hard, which would give me time to make new friends and still watch birds every day.

I became friends with Maus and Tim, who didn't watch birds themselves but found my hobby kind of amusing. When we were sneakily having a drink in a park somewhere, I could always point to something cool, like a European Green Woodpecker on the short-cut grass of the rosarium, a Song Thrush in the old willow tree opposite our favorite bench, or a Tawny Owl calling loudly at night when we were sitting in the park around the corner. Soon, seeing as many different species as possible became a popular pastime.

Gradually, I received some appreciation for my hobby.

I had a huge attic room with a roof terrace attached to it, the nicest room in the house; a perk of being an only child, I guess. Obviously, the attic was a popular place to hang out after school. The door featured my garden list, a record of all the species that I had seen in or above our garden since I was ten years old. There are now more than 150 different species on this list. When we were sitting on my roof terrace, my scope was always set up and ready. When a Great Spotted Woodpecker or starling landed in the top of a distant tree, I would immediately aim it at these birds, to the great disappointment of Maus, who might have been busy studying the breasts of my sunbathing neighbor, magnified sixty times.

After such boozy evenings, I always had to remove species such as "ostrich," "Mister Owl," or "Big Bird" from my garden list. I was disappointed that my beautiful list had been distorted, but I was nonetheless a little proud: Despite their lame jokes, my friends had shown a fledgling interest in birds.

Kiunga, Papua New Guinea (2016)

With its documentary series *Planet Earth*, the BBC made the birds-of-paradise famous in one fell swoop. In this series, David Attenborough introduces viewers to some of the nearly forty species found in New Guinea. Each and every one of them are dazzlingly beautiful birds, and together they feature all the colors of the rainbow. While showing off their decorative feathers, they assume the most amazing poses during their mating dance. After watching this series, even my nonbirding friends can understand where my fascination for New Guinea comes from.

The birds-of-paradise have been able to evolve into the most beautiful and diverse bird family in the world through sexual selection. The male with the brightest colors, longest decorative feathers, and most spectacular courtship is chosen by females for mating and is ultimately allowed to pass on his genes. Natural selection, first described independently in the nineteenth century by Charles Darwin and Alfred Russel Wallace, shows that animals that best adapt to their environment have the greatest chance of reproducing. Sexual selection can be compared with a group of teenage boys at a school dance, all trying to impress the prettiest girl in the class. Picky as she is, the girl will go for the smartest, most handsome, and best dressed boy, showing off his best dance moves.

With the birds-of-paradise, this process has taken on extreme forms. Generally, birds with too much plumage and too much outward show easily fall prey to monkeys, cats, martens, and other predators. However, these mammals do not occur in New Guinea, so this process of sexual selection has continued unbridled over a period of millions of years. With each new generation, the colors become a bit brighter, the feathers a bit longer, and the courtship a bit more exuberant. And the result is impressive.

For many birders, watching displaying birds-of-paradise in New Guinea would be the ultimate nature experience. However, only a few get the chance to experience this spectacle in real life, as New Guinea is far away and traveling to this destination is incredibly expensive.

During the flight to Kiunga, a remote town along the Fly River, deep in the interior of New Guinea, I look out of the window and see nothing but greenery. The rainforest around the cities is being cleared at a rapid pace, but due to a lack of infrastructure, the hinterlands remain largely untouched.

This country is by far the riskiest part of my Big Year, as my visit falls in the middle of the rainy season. And in New Guinea, that means landslides, muddy trails, flooded roads, and canceled flights. However, it is an El Niño year, and while some countries have to deal with extreme rainfall, New Guinea is struggling with persistent drought.

Once every few years, El Niño influences the climate in large parts of the world. Usually, there is cold water off the west coast of South America and warm water around New Guinea and Indonesia, and the trade winds over the Pacific blow from east to west around the equator. As a result, sun-warmed surface water accumulates around New Guinea and Indonesia. However, during an El Niño year, the winds turn and blow from west to east instead of from east to west and blow the warm water back to South America. There will be much less rain than usual in New Guinea and Indonesia, and much more precipitation in western South America. The nutrient-rich, cold water off the coast of South America warms up a few degrees, which means that fish stock is temporarily reduced, which is a disaster for people whose livelihoods depend on fishing.

Until just two weeks before our visit to Kiunga, it seems that we will have to cancel the three-day trip to the lowland rainforest along the Fly River: Due to prolonged drought in the mountains, there is not enough water in the river to supply the city. This crisis even makes the news in the Netherlands. The shelves in Kiunga's supermarkets have been empty for months, and famine and anarchy are imminent.

Noah had the same problem during his Big Year: Kiunga was temporarily out of reach, forcing him to abandon the idea of visiting the very species-rich tropical lowland rainforest of New Guinea. If I did manage to visit the lowlands of the Fly River, I would be at a huge advantage. Furthermore, I wouldn't want to miss this unique, very remote place for anything in the world.

Just before arriving in Port Moresby, I receive heaven-sent news via an email from our guide, Samuel: "The rain has finally arrived and there is enough water in the river again. Your trip to the Ekame Lodge can go ahead."

Kiunga is the New Guinean equivalent of a town in the Wild West: a couple of wide pothole-littered roads whose cheap Chinese asphalt has long since eroded, lined with small bars, shops, and dilapidated wooden hotels with boarded-up windows. Although it is still morning, every now and then, we see a drunkard stumble by. The road is riddled with red spots, visible remnants of the juice of betel nuts. Chewing the nut of the betel palm counteracts the feeling of hunger and has a slightly invigorating effect. When the nut is chewed, the red-colored residue is spat out. A red, toothless mouth is seen as a status symbol by many Papuans. Betel nut has been used for thousands of years and plays an important social, cultural, and sometimes even religious role in some communities. The lives of many Indigenous people revolve around chewing; once they have received their monthly salary, they immediately spend it on betel nuts and booze.

Samuel has snow-white teeth. "I see the effect chewing betel nuts has on people, which is why I decided to never start with it."

After we have loaded two barrels of petrol, a tray of water bottles, and a mountain of instant noodles into a motorized canoe, our long boat trip to the Ekame Lodge can start.

The first miles, we see hardly any birds as we sail at high speed in the middle of the river, but after a while we make a sharp bend to the right and we turn into a smaller tributary. It's like traveling back in time: Along both banks, we see a wall of impenetrable greenery, and every now and then, huge flying foxes and loudly screaming cockatoos fly overhead. The only people we see are passing by in dugout canoes, dressed in only skirts of dried banana leaves. When we wave at them, they look at us with curious eyes and give us shy waves back.

The Ekame Lodge turns out to be nothing more than a dilapidated hut on stilts. The menu for breakfast, lunch, and dinner consists of noodles with ketchup and sardines. We eat this culinary delight on the veranda, while we are attacked by thousands of mosquitoes.

"What a lovely place this is," I say.

Max looks at me in disbelief. "You must be out of your mind."

———

After dinner, we sail a few miles upstream. Armed with a flashlight, I sit in the foredeck and warn Samuel of dead branches sticking out of the water. After a while, he turns off the engine and lets the canoe bob downstream, making occasional adjustments with a wooden paddle. It is pitch dark and the crescent moon reflects in the black water. Above us twinkles the clearest starry sky I've ever seen, and now and then we hear the hollow song of a Papuan Frogmouth. Suddenly we hear a deep, foghorn-like "*hoooooeeeeeeeeemp*."

"A Forest Bittern," Samuel whispers.

The ghostly call is repeated every ten seconds, and only when we are nearly back at the lodge does the sound dissolve into the croaking of frogs and the chirping of countless crickets.

———

The next day we see a Twelve-wired Bird-of-paradise displaying on a bare branch, a fiery red male King Bird-of-paradise and a group of three Sclater's Crowned Pigeons. Max can't help but agree with me: "This is a great place indeed."

Mount Hagen, Papua New Guinea

I had booked a transfer to the lodge a few months ago, but when we walk out of the arrivals hall, our transport is nowhere to be seen. Since we are in New Guinea and not in France, we had been strongly recommended to refrain from walking out of the terminal to look for a taxi on our own. In doing so, many a tourist has been robbed of his luggage or, worse, his life. The unpaved streets behind the tall fence that separates the airport from the chaos of the city look far from inviting. Brusque-looking men stand behind the fence with bare chests, yelling all kinds of things at us. I feel like we've arrived in a real-life version of *Prison Break*.

We are here to visit the world-famous Kumul Lodge. Located at an altitude of 3,000 meters in a cloud forest, this lodge owes its fame to a 3-yards-long, moss-covered bird feeder table. Every morning, this table is filled with pieces of papaya, mango, and banana. Because this is New Guinea, it does not attract tits, finches, and sparrows, but parrots and birds-of-paradise.

We see a friendly looking woman at the baggage carousel, and we decide it would be safe to approach her.

"Do you know of a trustworthy person who would be willing to give us a lift to the Kumul Lodge?"

The woman nods and picks up her phone. Her "husband" is a security guard and he might want to take us in his pickup truck. For a few minutes, she's speaking frantically on the phone, after which she states without any embarrassment, "My husband can take you to the lodge. For two hundred dollars."

Two hundred dollars? She clearly knows we depend on her.

Moments later, a black pickup truck pulls into the parking lot. The windows are protected by bars of reinforced steel; the giant vehicle reminds me of something out of *The A-Team*. The fierce-looking Papuan getting out of the car could have played the part of B. A. Baracus. His upper arms are as thick as my head, and he looks at us like a hyena about to tear up a baby gazelle. When he shakes my hand, I think he can easily yank my arm off my shoulder.

"I am Michael and I am taking you to the Kumul Lodge."

In my mind I hear: *I'm Michael and I'm going to rob you.*

Michael's companion, who has come with him for no apparent reason, has an equally dodgy appearance.

"That must be the one who will soon dump us in a ditch somewhere," I say to Max in Dutch.

He laughs out the other side of his face.

After we negotiated the price down to $160, we leave for Kumul Lodge. I'm sitting in the passenger seat next to Michael, and Max sits in the back next to Michael's companion, who introduced himself to us as Christopher.

Michael and Christopher are engaged in a frantic discussion, arguing in a local dialect, and we think we frequently hear the words "money" and "bags." I get a horrible feeling in my belly.

As we leave Mount Hagen and drive into the mountains, I feverishly look around for a sign with KUMUL LODGE. Will this winding road end at the lodge or at the ditch I was joking about an hour ago?

I plug a cable into my phone, and a few seconds later the pumping beats of Dr. Dre resound from Michael's subwoofer. Meanwhile, I lie and pretend that Max and I have been kickboxing since we were ten. I show him the scar on my right upper arm, which I did not get from kickboxing but from a drunken night out. I'm like a bird that puts up its feathers to appear bigger and more dangerous so as not to fall prey to a hungry predator.

My bluff seems to have the desired effect. Michael nods approvingly when he hears the hip-hop and sees my dramatic scar. Christopher smiles broadly, his betel-eaten gums almost making us heave.

The atmosphere in the car remains excruciating, and I breathe a sigh of relief when I see the sign KUMUL LODGE 10 KILOMETERS along the road. But we will not get rid of them that easily.

"You'd better give us a tip," Michael states with a wicked grimace, making a painful-looking fist with his right hand. "We now know where you are staying."

I decide our life is worth more than forty dollars and hand him the last of my cash.

Our fears of death are immediately gone when we see a familiar face on the driveway of the lodge.

"Guys, I'm so glad to see you," Vincent says.

Five days ago he texted me that his flight to Port Moresby had been canceled. He was stranded in Hong Kong.

I FEEL LIKE SCREAMING MY HEAD OFF, was his last message.

As we carry our things to the room, he tells us that it took him three full days to get here. Three days of travel out of ten vacation days—what bad luck. Max and I had been on more than thirty flights and only one of them had been delayed.

We go straight to the bird feeder table. However, it is low season and we are the first guests to stay here in months. The table has not been provided with fresh fruit for ages, and birds are nowhere to be seen.

But as soon as a fresh load of papayas and bananas is brought out and thrown on the table, the forest comes alive. Suddenly, a large dark bird with a long tail, a fine black-and-white banding on the belly, and a wafer-thin, sickle-shaped beak appears from the right: a female Brown Sicklebill. This bird-of-paradise is one of the largest passerines in the world—the females grow to over a foot and a half from beak to tail tip, and the males can reach a size of more than a yard. She lands on the table and throws herself on the papaya. She grabs the pieces of fruit between the tips of both mandibles and tosses them into the back of her throat in one smooth movement.

Then a black bird with a green forehead flies in. Behind it, we see a fluttering snow-white ribbon of more than a foot and half in length. It's a Ribbon-tailed Astrapia, the bird with the longest tail compared to its body. This young male already has an impressive appearance, but when you consider that an adult bird's tail grows to over a yard in length, he still has a long way to go.

Then the Common Smoky Honeyeater makes his entrance: a black songbird with bare, bright yellow skin around its eyes. While he furiously pecks at a banana, he receives a blow from the sicklebill. The honeyeater leaps to the side, and within seconds the bare skin around his eyes transforms from bright yellow to bright red. It looks unreal. I had read that there are bird species that do this, but I had never seen it with my own eyes. When stressed, the bird pumps blood into the thin capillaries of its face at a rapid rate. With this red color, the honeyeater gives off a signal of aggression to scare off the other candidates on the bird feeder table.

Meanwhile, Vincent lights a cigarette. I look at him.

"Haven't you stopped yet?"

"You know what? If you discover a Blue-capped Ifrit, a Tit Berry-pecker, and a Wattled Ploughbill for me, I will stop smoking."

All three species belong to bird families that occur only on New Guinea, and it's no surprise that they are at the top of Vincent's wish list (7 of the more than 240 bird families occur exclusively on New Guinea).

He has barely finished his sentence when I see a yellow-green bird with a black head and bright yellow cheeks land on a bromeliad. A Tit Berrypecker! Bizarrely, this bird bears a striking resemblance to our Great Tit, a fine example of convergent evolution—the evolution of the same functions or external features in different, unrelated species.

The next morning, when we take a seat at the bird feeder table again, I see an ocher-colored bird with a bright blue cap, crawling like a nuthatch along a mossy trunk behind the table. There can be no doubt, a Blue-capped Ifrit! This is one of the very few venomous bird species. Handling one can cause numbness, tingling, and a burning sensation. The name *Ifrit* is derived from Arabic mythology: a super-natural being that symbolizes the element of fire.

"And that's two," I say triumphantly.

Vincent nervously lights another cigarette, knowing that this might just be his last one.

Number three, the Wattled Ploughbill, is a lot rarer, but I'm abso-lutely determined to get Vincent off his addiction once and for all.

As we walk the forest trails around the lodge, I keep my ears open for the high, tenuous whistle of the ploughbill, however without success. At the end of the afternoon, I decide to try my luck one more time in a valley along a steep forest path, nearby the lodge. I hear a sound that is strongly reminiscent of a dog whistle, and suddenly I see a moss-green bird flying over. The bizarre, bright pink wattles at the base of its beak tell me unmistakably that it's a Wattled Ploughbill.

I run as fast as I can back up the steep path, but that proves to be quite a challenge in the thin mountain air. My breathing gets heavier with every step, and I almost slip a few times. When I reach the lodge my legs are sore, and I arrive gasping for breath at the bird feeder table, where Vincent and Max are relaxing and enjoying a Coke.

"Ploughbill ... All the way down ... Come on," I gasp. Sweat pours from my body.

Vincent quickly takes three more deep draws and blows the smoke through his nose with a blissful look in his eyes. Then he puts the cigarette out in an ashtray.

Moments later, the three of us are looking at the ploughbill, who, luckily, has remained in the same spot along the path.

"Unbelievable," Vincent says, pulling an electronic cigarette from his pocket. "A hundred times better than smoking."

"You can't be serious!"

"What? This is just water vapor with a taste."

I look at him in astonishment and then focus my binoculars back on the ploughbill. Was I just imagining it, or did I just see his pink wattles shake in disapproval?

Port Moresby, Papua New Guinea

In mid-March I say goodbye to Vincent and Max in Port Moresby. They will fly back to the Netherlands, and I will continue to Australia. We scored great in Papua New Guinea, better than I ever thought possible, especially given the time of year. I have now more than made up for my disadvantage compared to Noah, and I have full confidence again.

After Max and I have lived like a married couple and shared joys and sorrows for two and a half months, it's time to say goodbye. He was a terrific travel companion, and it will take some getting used to to continue on my own. I wrap my arms around him. "I'm going to miss you, dude." Saying goodbye to Vincent is also hard. He will always be a mentor to me, that tough guy who listened to grunge music, smoked cigarettes, talked about women, and threw my backpack between the sea buckthorns at the bottom of The Volcano when I talked too much, but who also taught me the art of birding.

The next twelve days are the most ambitious of my entire year: Australia, Tasmania, and New Zealand in record speed, with twelve domestic flights all connecting seamlessly. If all goes well, I will lay the foundation for a new world record during this part of the journey. If things go wrong, I will probably be at home in two weeks, laying on the couch with burnout.

Adelaide, Australia

Just before sunrise, I am woken up by the deep, foghorn-like song of an Emu. What a great sound to wake up to. Moments like these make me realize how incredibly privileged I am to be able to undertake this journey.

My guide and fellow Dutchman, Peter, and I camped in the middle of the bush last night so we could start birding with the first light of dawn. In the span of one morning, I hope to see all the special bird species of the dry mallee forest, which consists of low, thorny shrubs and various eucalyptus species. And there are quite a few.

I am particularly keen to see the Black-eared Miner. About a hundred years ago, this species led a carefree existence in the vast mallee forests of South Australia. From the 1930s, this type of forest was rapidly cleared to make way for agriculture. The opportunistic (and closely related) Yellow-throated Miner took advantage of this and moved into the Black-eared Miner's habitat. Both species soon began interbreeding, resulting in fertile offspring. At school, we were taught that two animals belong to the same species only if they can have fertile offspring, however that is an outdated concept. Sometimes, different species within the same genus may also produce fertile offspring, such as the Tufted Duck and the Greater Scaup.

Today, there are only a few hundred genetically pure Black-eared Miners left. If you are lucky enough to encounter a flock of miners in Gluepot, Australia, there is a good chance that it consists of only Yellow-throated Miners and (back) crosses. Obviously I want to do everything in my power to lay my eyes on this disappearing species.

Our day got off to a good start. The Emu that woke us up with its hollow song scurries away as soon as I walk out of the tent. The Emu is the second heaviest bird in the world after the ostrich and is a species that cannot be missed during a Big Year, among other giants such as the Andean Condor, the Wandering Albatross, and the Great Bustard.

In the course of the morning, we see an impressive list of species, but the Black-eared Miner seems to be untraceable. By eleven, when the sun starts to sear, we throw in the towel with pain in our hearts. Peter seems to find missing this species even worse than I do; for him it is a matter of principle to give his compatriot a maximum score.

When we drive out of Gluepot, I doze off in the passenger seat. I abruptly wake up as the car comes to a halt with a jolt. We didn't hit a kangaroo, did we?

"A group of miners flew across the road right in front of us," Peter says excitedly.

We jump out of the car and run into the bush. It doesn't take long before we get to see them. Between the Yellow-throated Miners and crosses, we see two pure Black-eared Miners, recognizable by their dark rump and the black markings around their eyes.

The countdown for this species seems to have begun. Will these specific birds themselves negatively be affected? Probably not. Slowly but surely their gene pool will merge with that of the Yellow-throated Miner. I find it very worrying that we have been able to wipe out an evolutionary process, spanning many tens of thousands of years, in less than a century.

Wellington, New Zealand

The chronic sleep deprivation is now starting to play tricks on me. When I walk into the bathroom at the airport to splash some water on my face, the person in the mirror scares me. I have deep bags under my bloodshot eyes and gray skin from fatigue. The grubby and bushy beard even makes me look a bit like a Taliban fighter.

"This is what you wanted, so be a man," I say to myself. I slap my cheek with my flat hand in order to wake myself up.

When the plane lands, it is just getting dark. Great, that means I can go to bed quickly and get no less than seven hours of sleep. But that dream is shattered when I walk into the arrivals hall with my backpack, where I am greeted by Johannes.

"Hey man, are you ready to go and look for some kiwis?"

Did I understand that correctly? Did he really say *looking for kiwis*? All five kiwi species are shy, rare, and endangered. Since we're at an airport next to New Zealand's capital, it seems highly unlikely that we'll see one of them.

"Just wait and see. In an hour you'll be watching a Little Spotted Kiwi."

I know Johannes from the Netherlands, where he was one of the more fanatical young birders, just like me. He moved to New Zealand for his studies and has now become *the* expert on the South Georgia Diving Petrel, a seabird the size of a starling that resembles a flying mini-penguin. For several months a year he lives all by himself on the inhospitable Codfish Island, studying a small, remote colony of these birds.

As soon as we are in the taxi, Johannes asks me to give him my shoes. Somewhat hesitantly, I accede to this strange request. He takes out a pair of tweezers and begins to remove mud, grass, and seeds from the soles and laces with surgical precision. The ritual resembles a baboon mother grooming her young. While doing this, he talks about the revolutionary conservation project that we are about to visit.

Zealandia is located on a small peninsula and is separated from the rest of the North Island by a concrete wall. On this peninsula, nature looks like it did in New Zealand thousands of years ago.

Through millions of years of isolation and the lack of land mammals—only two primitive bat species, which can barely fly, are found in New Zealand—birds have taken up all the niches in this ecosystem. Before the Polynesians first set foot here in the thirteenth century, at least eleven species of moa lived here. Moa were enormous ratites, sometimes weighing up to 440 pounds and growing 9 feet tall. Furthermore, the island was inhabited by dozens of other unique bird species, including an eagle with a wingspan of almost 9 feet that hunted moas. Within a few centuries, the moas, and consequently also the eagle, all went extinct.

When Abel Tasman "discovered" New Zealand in 1642, it spelled the end for many more bird species, as the Western settlers brought a large number of invasive animals and plants, such as cats, foxes, mice, rats, possums, all kinds of European songbirds, and recently even wasps. The unique local fauna and flora lost out, and the animal and plant life of today is only a shadow of what it once was.

"That's why I clean your shoes so thoroughly," Johannes says. "I don't want some grass species from New Guinea to ruin things even further."

Twenty years ago, a rescue operation was launched to save the last remaining native fauna from extinction. Offshore islands were cleared of the most aggressive introduced predators, such as rats, mice, cats, stoats, and foxes. Subsequently, all kinds of endangered native bird species were moved from the mainland to these islands. The government wants New Zealand to eventually become completely free of land mammals—except for the two bat species—to give the original flora and fauna free rein again. Zealandia is one of the first sanctuaries on the mainland where all non-native mammals have been eradicated.

Where we humans have destroyed nature the most, we also work the hardest to rescue it. I think that's part of human nature: We always have to first fail disastrously before we learn from our mistakes.

———————

Exactly an hour after we left the airport, we indeed find ourselves in a pitch-dark forest watching a Little Spotted Kiwi scurrying around in the ghostly red glow of Johannes's flashlight. The bird has powerful legs, which it uses to kick leaves and branches to the side. Its wings and tail are so underdeveloped that they are no longer visible, and its body is covered in short, fluffy feathers, making it look like a kind of walking fluff bundle. Every now and then it pulls a worm out of the forest soil with its long beak. No wonder kiwis have specialized in this type of food: There are nearly 200 species of worms in New Zealand, of which—you guessed it—no fewer than 23 have been introduced by humans.

When I finally go to bed, I'm exhausted. The clock on my phone reads half past two; in less than five hours my alarm will go off again. Despite this prospect, I fall asleep with a smile on my face. I just saw a kiwi, and no one will ever be able to take that away from me.

Perth, Australia

My plan is to drive a rental car from Perth to Cheynes Beach and to return in two days, a journey of more than 700 miles. Madness, of course, but this reserve is one of the last strongholds for three bird species: the Noisy Scrubbird, the Western Bristlebird and the Black-throated Whipbird, collectively known as "the big three." They have three things in common: they are endangered, rare, and notoriously hard to find.

Ross, a cheerful Australian birder with a bushy mustache, is driving the rental car. I sit next to him on the passenger seat with the seat kicked back. We left Perth in the middle of the night, which means that Ross will have to cover the first 190 miles in the dark, right through the outback. This in itself would not be a problem, were it not for the fact that kangaroos tend to jump onto the road out of nowhere.

"You better get some sleep, mate. I'll keep an eye on the road."

Falling asleep without a care in the world proves to be difficult. Yet, exhaustion soon takes the better of me, and it doesn't take long before I close my eyes.

I wake up to the typical *psssst!* sound of a can being pulled open by its pop-top. Ross is addicted to Pepsi. The floor of the car is littered with empty cans, and two trays with cans are still waiting on the back seat. I also open a can out of solidarity. My still-empty stomach contracts after the first sip of the sugary sweet chemical substance.

A thin strip of orange light on the horizon reveals that morning will break soon. Slowly the contours of the landscape become visible. We are driving on an endless dirt road and alternately pass barren plains and dry forest.

When, a few hours and ten cans of Pepsi later, we arrive in Cheynes Beach, we have already gathered an impressive list of species.

Ross is tireless—which is not surprising given the high level of caffeine in his blood—and suggests looking for the big three right away.

"Hurry up, mate. We're on a date with the Noisy Scrubbird."

I've barely gotten out of the car when Ross is already bouncing through the heather like a gazelle, his camera dangling from his shoulder.

The Noisy Scrubbird was described as a new species in the mid-1800s, but a little over half a century later, the species was considered extinct due to extensive land clearing and predation by feral predators such as cats and foxes.

In the early 1960s, the scrubbird was rediscovered near Two Peoples Bay, in a tiny remaining patch of heather, its favorite habitat. The population consisted of no more than fifty specimens. From that moment on, the species was intensively monitored and protected. Conservationists declared war on foxes and stray cats, and the natural environment was restored where possible. In addition, a few scrubbirds were captured and released in new, suitable places.

As a result of these conservation measures, between 1,000 and 1,500 Noisy Scrubbirds are now living in the dense undergrowth of southwestern Australia. This is undoubtedly a success. Still, the future of this species hangs in the balance, as the current population stems from no more than fifty birds. This means the genetic variation is limited, and the population is prone to inbreeding. The number of heath fires has also sharply increased due to persistent drought and rising temperatures as a result of climate change.

We manage to see the Western Bristlebird and the Black-throated Whipbird that afternoon, but the Noisy Scrubbird lives up to its reputation and remains invisible, although we hear the call of several specimens. Therefore, it does count for my Big Year. But that does not alter the fact that I really want to see this species, with its eventful history. Fortunately, we have another opportunity tomorrow morning.

That evening, Ross and I are sitting in front of our cottage on the campground. There is no wind, and the sun slowly sinks behind the heather-covered hills. I am having a beer and Ross a glass of Pepsi with Jack Daniel's, while a Brush Bronzewing—a pigeon with beautiful yellow, green, and purple shiny feathers on its wings—is scurrying through the grass in front of us. I tentatively ask Ross if he is not concerned about his health, drinking so much Pepsi, but his answer is no.

"No, mate. I'm on a diet. I used to drink twenty cans of Coke a day. It contains sugar and other junk, so now I only drink Pepsi Max."

The next morning we continue our search for the Noisy Scrubbird. While we're walking on a narrow path through the heather, we suddenly hear his unmistakable, explosive, nightingale-like song right next to us. A few seconds later, a dark brown bird with a bushy tail runs across the path just in front of us. In the blink of an eye, he disappears again into the dense vegetation. It wasn't a world-class sighting, but considering the reputation of this species, I count myself lucky.

"Psssst! Pssssst!"

Ross has opened two cans to celebrate the occasion. I give him a high five and together we toast to the Noisy Scrubbird.

Amsterdam, Netherlands (2014)

Sometime in the summer, the phone rang: Do I want to participate in a game show about birding? "Do you mean a kind of *Expedition Robinson*, but for birders?"

A week later I was invited to the Eyeworks office in Amsterdam, where I was told that this was indeed the essence of the television show *Under the Spell of the Condor*. By linking birders to famous Dutch people, the show makers wanted to introduce the Netherlands to the wonderful hobby of "bird spotting."

Bird spotting. Upon hearing that term I had to suppress a gag reflex. I somehow always associated the term *spotting* with something silly. It conjured up an image of a sunburnt tourist in a totally over-the-top khaki safari outfit with mini binoculars around his neck with the lens caps still on, who thought he could see tigers in Africa and wasn't sure if a penguin was a bird or a mammal.

I was definitely not a fan of game shows, but the idea of making my hobby more popular with a wider audience—starting with the elimination of the term *bird spotting*—naturally appealed to me.

In the fall of 2014, the time had come, and the recordings for *Under the Spell of the Condor* started. For the first episode, we went to Texel. I was paired up with television host Gallyon. We competed against three other birder-celebrity teams by completing assignments related to bird identification and bird photography. Gallyon was extremely fanatic, just like me. We won that first episode quite easily, and together with two other teams we went through to the next round in De Biesbosch National Park.

While we were sitting around the campfire in front of our tents in the Biesbosch, the presenter, Tooske, said to me, "I would like to introduce you to the new team: actor Mike and the beautiful biologist Camilla Dreef."

I turned around and instantly fell in love.

"Hi, I'm Camilla," said the beautiful biologist, shaking my hand.

"Arjan," I said with a wide grin on my face.

"Again!" shouted the director, who immediately realized that this could be the beginning of a fledgling romance.

Camilla and I shook hands about ten times, and each time it was captured from five different camera angles and even with a drone. Each time I shook hands, I tried to learn more about Camilla, much to the director's dismay.

"If you say something different every time, I can't possibly put these images together! Do you know how much a minute of recording time costs?"

I didn't care at all how much a minute of recording time costs, and at the tenth handshake I asked Camilla how her interest in birds had started.

It turned out that she was doing research on spoonbills at the University of Amsterdam. Wow, I was in awe; that was something well beyond my half-finished study of archaeology.

That evening, as we were drinking wine around the campfire, I heard the call of a Barn Owl. For Camilla, this was still a new species. I explained to her that you could hear distant sounds better by holding your palms behind your ears like saucers.

So we stood side by side listening to the hoarse, ghostly cry of a Barn Owl, both with our hands behind our ears. Of course, this moment was captured on film from a hidden position.

In episode four, Gallyon and I dropped out of the show because we didn't agree with a new game element that involved making a team swap with the two newest candidates. Camilla and Mike continued and eventually made it to the final in Peru, where they saw the Andean Condor, the world's second-largest flying bird, after which the show was named.

Camilla and I continued to see each other after the show ended, and a year later, when *Under the Spell of the Condor* finally aired and we'd been together for over a year, 600,000 people were able to enjoy the moments of our first meeting.

Schiphol, Netherlands (2016)

After a crazy twelve-day tour through Australia and New Zealand and a long flight via Dubai to Amsterdam, I arrive at Schiphol in the afternoon of April 1. The hardships of the last few weeks have been worth it: After exactly three months, my count stands at 2,060 species. I am now well ahead of Noah's pace, while I haven't even started my journey through Africa and, last but not least, the most bird-rich continent of all, South America. Slowly but surely, I start to realize that breaking the world record could become a reality. But everything will have to go well. Falling ill, for example, will not be an option.

My lightning visit to the Netherlands turns out to be the right move, especially for morale. In the evening, I thoroughly enjoy catching up with my parents and Camilla. I have the opportunity to wash all my clothes properly and to repack my bag. After three months of nonstop travel, I know exactly what is unnecessary and what is lacking. As I carry only hand luggage, I'm very limited in my choices. I exchange my heavy, clumsy poncho for a light rain jacket. I didn't use two of the three hard drives I'd carried, so they can stay at home. I stuff some extra socks in the side pockets of my backpack, an extra pillowcase between my T-shirts, and a new supply of malaria pills in my toilet bag.

It is wonderful to sleep with Camilla in my own bed for two nights. During the first night she wakes me up because she hears an owl.

"Do you hear that in the distance?"

A few seconds later, I also hear the unmistakable "*Hoouuuh . . . Ho ho hooooouuuh!*" of a male Tawny Owl. In retrospect this was the only Tawny Owl that I would observe during the whole year.

When, two days later, I say goodbye to Camilla and my parents in the airport and have to board a plane once again, I'm more than ready for round two of my Big Year.

Addis Ababa, Ethiopia

While Africa has rapidly modernized, Ethiopia has remained largely traditional. The Western world has had a lot less influence here than in other African countries, and this is noticeable in almost all facets of daily life. This is reflected in their traditional way of living, dressing, and cooking, and in their music.

The Ethiopian Highlands are bisected by the Great Rift Valley: an elongated rift valley that is 100 miles wide in some places, with gorges that can reach depths of thousands of meters and mountain peaks that reach above 4,500 meters. That's why Ethiopia is rightly called the roof of Africa.

The Ethiopian Highlands were formed about 75 million years ago. At that time, the Horn of Africa and the Arabian Peninsula were still connected. The pressure of magma pushed the earth's surface upward over a period of tens of millions of years. The area in between was pushed down, which created the Great Rift Valley, the Red Sea, and the Gulf of Aden. The higher area was divided into three parts, and so the highlands of southwestern Arabia and the northwestern and southeastern parts of the Ethiopian Highlands came into being. The birdlife in these mountain ranges reflect this: The avifauna of the southwestern part of the Arabian Peninsula bears a striking resemblance to that of the Horn of Africa.

Millions of years of geographic isolation have allowed the plants and animals of this highlands region to develop in a unique way. You could say that this area is a kind of onshore island. It is the exclusive habitat to dozens of bird species. Therefore, this area is an absolute must for my Big Year. And not unimportantly, Noah decided to skip this top destination.

When I step off the plane in Addis Ababa, I immediately smell the typical African scent of grass, smoldering fires, and dry earth. I'm met by Merid, a small Ethiopian man with huge dreadlocks who could be a long-lost son of Bob Marley. I learn that Merid's haircut is an aesthetic choice, as he is not a Rastafarian himself. I throw my bag in the back of a 4×4 and we leave for Debre Libanos.

On our way there, we stop for a traditional Ethiopian cup of coffee. We take a seat on brightly colored floor cushions around a low, round table. The coffee is prepared over an open fire, while all kinds of spices are added. The restaurant soon fills with an aromatic, nutty scent. I take a sip and experience an explosion of taste: This fragrant, spicy cup of coffee is nothing like the uninspired dishwater that is sometimes served in the Netherlands.

With the coffee comes a sour pancake, called injera, which is never missing from an Ethiopian meal. Merid shows me how you are supposed to eat such a pancake. He tears off a piece with his fingers and dips it in a lentil sauce. When I follow his example, I am pleasantly surprised. However, I now understand why almost every tourist who visits Ethiopia gets diarrhea: If your table companions don't wash their hands, you're out of luck.

Debre Libanos, a thirteenth-century monastery, offers an overwhelmingly beautiful view over the Great Rift Valley. As I carry my bag to my room, a gigantic Bearded Vulture circles up from the valley at a distance of less than 20 yards. With a long, wedge-shaped tail and slender wings, it effortlessly gains height on an invisible thermal bubble rising up from the valley. It comes so close that I can even see its pale iris and the black, brush-shaped feathers on the base of its beak. Bearded Vultures breed on inhospitable, vertical cliffs, where usually there is no water. Those small feathers at the base of their beak allow them to carry drinking water for their young.

A large troop of geladas—baboon-like monkeys with waving manes and long, dangerous-looking fangs—graze like a herd of sheep on a windswept, grassy slope. This is the only monkey species that grazes. They occur exclusively here and are the last living descendants of *Theropithecus*, a genus of prehistoric apes that used to live in large sections of Africa and southern Europe until the Pleistocene.

This monkey is not the only prehistoric relic in the Ethiopian Highlands. Dozens of animal and plant species that live here are the closest relatives of species that died out millions of years ago. This makes this part of Africa extremely interesting for evolutionary biologists, as well as for archaeologists and anthropologists, because this is where the cradle of humanity lies. It was not far from here that they found the 3.2-million-year-old fossilized skeleton of Lucy—named after the Beatles' "Lucy in the Sky with Diamonds"—the most complete skeleton of the early hominid *Australopithecus afarensis* ever found. Thanks to this discovery, we know more about the evolutionary path that humans have followed.

During my first two days in Ethiopia, I see no less than 127 new species. Optimistic as I am, I assume that I will easily break the record at this pace. At this point, I don't yet realize that I'm making a critical miscalculation.

Bale Mountains National Park, Oromia, Ethiopia

The road meanders up through a fairytale cloud forest. Long tufts of silver-gray beard moss cover the branches of ancient trees, waving gently in the wind. We regularly drive by herds of mountain nyalas, stately, gray-brown antelopes with long twisted horns that flee deeper into the forest when they see our 4×4. We are on our way to the Sanetti Plateau, a vast plateau at an altitude of more than 4,000 meters.

Once we have passed the tree line, the road flattens out and the landscape undergoes a total metamorphosis. The forest gives way to short grassland with heather, giant lobelias, and extremely tall phallus-shaped red flowers that tower over all other vegetation. Groups of Spot-breasted Lapwings walk along the banks of shallow ponds; these are elegant waders with bright yellow legs, a black-spotted chest, and a striking white eyebrow stripe.

This completely unique landscape is also home to the Ethiopian wolf, the rarest canid in the world.

Until the mid-nineteenth century, the Ethiopian wolf led a carefree existence on the vast plateaus of Ethiopia. It had no natural enemies and fed almost exclusively on giant mole rats, rodents that occur only in this area.

But the fertile soil made this area extremely suitable for agriculture, and in a relatively short period of time, much of their habitat was converted into arable land. Shepherds let their sheep, goats, and cows graze on the nutrient-rich grasses, resulting in overgrazing and erosion. However, the main danger for the Ethiopian wolf is a lot less obvious. The shepherds are invariably accompanied by dogs that carry contagious diseases such as distemper and rabies.

Now, only a few hundred Ethiopian wolves remain living in the wild. Scientists and conservationists are doing everything they can to save these beautiful animals from extinction through a vaccination program. You can imagine that this is far from easy. The vets camp on the plateau, at nighttime temperatures of 5°F, and have to inoculate all the animals in a pack, one at a time, with a stun gun. Each vaccine costs $200, so it takes $20,000 to vaccinate all the wolves living along the edge of the plateau. The aim is to create a buffer zone that should protect the remaining wolf population.

Holes are in the ground everywhere; they remind me of rabbit burrows.

"These are the dens of giant mole rats," Merid explains to me.

A little later we see a specimen. It is the size of a rat, with a thick head and extremely large front teeth, which it uses to excavate yards-long tunnels.

It strikes me that almost every mole rat is accompanied by a Moorland Chat, a gray-colored, robin-like bird. The chat keeps a close eye on the mole rat, and every now and then it pecks at a beetle or maggot that is tossed up by the burrowing rodent. It's a symbiotic relationship: The mole rat makes it easy for the chat to forage for food, and the chat warns the mole rat of danger—for example, a hungry Ethiopian wolf.

———————

At the end of the morning, we still haven't seen a wolf. This is not surprising, of course. I wistfully stare out the side window as we drive back to the edge of the plateau. Suddenly, I see an orange dot moving far away across the plain. Merid follows my gaze and hits the brakes. I can hardly believe my eyes, it's an Ethiopian wolf! His physique is somewhere between a wolf and a jackal, but he is unique for his strikingly long legs, shiny orange fur, and contrasting white throat and belly. Maybe he picked up the scent of a female somewhere on the other side of the plateau. Imperturbable, he runs up a hill, into an uncertain future. I follow him until he disappears behind the top.

The Hague, Netherlands (2015)

Six months before my Big Year, my grandmother became seriously ill.

"It's cancer," my mother told me. "Metastasized to her lungs and lymph nodes."

I had always been fond of my grandmother, and she was fond of me. When I was a little boy, we often went to a holiday cottage with Boomer, her golden retriever. Her dog's name was Boomer because my grandmother had actually wanted a boomer dog, which is a little lap dog that looks more like a guinea pig than a dog. When we walked the dog, I, of course, mainly looked at birds. My grandmother was there when I saw my first Yellowhammer singing on top of a hawthorn, and when I thought I had discovered the rare Arctic Redpoll in a field in Groningen. In hindsight, it obviously was just a big Mealy Redpoll.

My grandmother loved that her grandson was so interested in birds. "More children should do that, being out there in nature. Constantly sitting indoors only makes people stupid," she would say.

She came from a humble family of twelve siblings. Her mother died just before World War II, when she was seven years old. Until she was twelve, she, with many of her brothers and sisters, lived in an institution on the Veluwe.

Being the eldest daughter, she had to take over the role of her mother after the war, and so she didn't have much chance to get an education. That's why she thought it was so important that I worked hard at school and that I broaden my horizons. She had never had that opportunity herself.

"Make sure you see as much of the world as you can, because before you know it you'll be as old as me and then it will be too late," she told me.

Words I certainly took to heart.

───── ·────

My grandmother was visibly upset that she had become ill so close to my departure. What if she died during my trip?

She took my hand, and from her hospital bed looked at me with a determined look. "Whatever happens, don't interrupt that trip."

I nodded. "I promise."

But we both knew I would come back for her, of course.

───── ·────

In the end, she died just before I left, a few days before Christmas, as if she had arranged it that way. The whole family, my mother and I, my aunt and my nieces, we all were at her bedside when she released her last breath.

I gave a speech at her cremation, because that's what my grandmother would have wanted. I spoke of how important nature was to her and how she always encouraged me to make something out of my hobby. Because constantly sitting indoors only makes people stupid.

Mount Kenya, Kenya (2016)

In mid-April I drive with a motley crew in a rattling Toyota Land Cruiser along the foot of Mount Kenya, the second highest mountain in Africa. Behind the wheel is Zareck, a Kenyan of Indian origin. He has shoulder-length hair, a huge beard, and prefers to walk around barefoot. Joseph, a local guide from a small village on the shores of Lake Baringo, and Ethan, a young American who works as a guide for a South African birding company, sit in the back of the 4×4.

I met Ethan six months ago in the Netherlands. He had approached me via the Birding Pal website, a place where birders worldwide meet and can arrange to go birding together. He had a half-day stopover in Amsterdam and was looking for a birder who could show him some European bird species. From Schiphol, we drove straight to IJmuiden, and as we walked across the south pier I told him about my plans. He immediately suggested that he be my guide in South Africa and accompany me in Kenya.

The landscape gradually changes as we drive north. Last night we camped on the flank of Mount Kenya, surrounded by tropical mountain forest, and now we are driving through a dry landscape with acacias. Along the way, market vendors come and go, trying to sell all sorts of things: carvings, beaded necklaces, and brightly colored rugs and fabrics, but also slippers, shoes, clothes, secondhand car parts, and even entire double beds. We stop at a vegetable stand to stock up on food. Just like last night, we will be camping, and that means that I can once again practice my other hobby: cooking. Although our culinary indulgence will be limited to a minestrone soup prepared in an aluminum pan on a butane burner, I'm already looking forward to it. Apart from Camilla and my parents, cooking is the only thing I really miss during this year. As for the rest, my Big Year is such a succession of new experiences that there is simply no time to think of other things. World politics, for instance, completely passed by me for almost four months. After such a long period of intensive travel, it has become completely normal that my life revolves around only one piece of hand luggage, binoculars, a camera, and seeing as many bird species as I can.

A riverbed winds like a green lifeline through the otherwise bone-dry landscape of Samburu National Park in central Kenya. This oasis supports an explosion of life and contains large populations of elephants, lions, cheetahs, leopards, and rare ungulates, such as the beautifully marked reticulated giraffe, the white-bellied Grévy's zebra, and the gerenuk, an elegant antelope-like animal with a remarkably long neck that regularly stands on its hind legs to nibble juicy young leaves from acacia trees.

There is just as much to be experienced in terms of birdlife. It is one of the few places where one can see the Somali Ostrich, which differs from the Common Ostrich with its bright pink shins. Or the Vulturine Guineafowl, with its white-spotted blue plumage and long black-and-white decorative feathers. It is a distant relative of the guinea fowl that ends up on many plates on Christmas Eve.

Near the entrance of Samburu National Park, there is an abandoned camping site where we set up our tents. We have a beautiful view over the river, and while having a beer, we watch the sunset casting a red glow over the desolate landscape. When darkness sets in, I start to cook. Slender-tailed Nightjars fly around us, and in the distance we can hear a spotted hyena laughing. Cooking has always had a therapeutic effect on me; I completely relax after all the hustle and bustle of the last few weeks. How wonderful it is to concentrate on something other than traveling and bird watching.

However, our peace is suddenly disturbed by four cars that, blasting out a thumping bass, drive onto the camping site. Twelve drunken Kenyans get out of the cars and demonstratively choose their pitch within 5 yards of us, while the camping site covers several hundred square yards and there's not another living soul to be found. Judging by their short skirts, fluorescent pink lipstick, and deep cleavage, the ladies have been paid by the men for their company. They put on loud music and the group starts drinking.

Zareck shakes his head. "This could become a troublesome night."

His suspicion turns out to be right, as a little later, when the soup is ready, the party next to us has degenerated into the kind of party where the men sit on plastic chairs and the ladies dance in front of them, pushing their buttocks almost in the faces of the men. Moreover, it seems that this party has surreptitiously moved into our direction by a yard or two. Ethan, who is sitting closest to the group, can almost feel the breath of one of the Kenyan ladies on his neck.

"Come and sit on mommy's lap with your *skinny ass*," the lady says while she tries to grab his behind, which is remarkably skinny indeed.

Ethan looks at her through his glasses, dumbfounded, and moves his chair a few feet away.

It strikes me that the men are telling tall tales in English, and not in the more common Swahili, as if they're trying to impress us.

While I'm dishing up the soup, one of the ladies—with a not-so-skinny ass—jumps onto one of the men's laps, who, with a lot of hullabaloo, falls through his plastic chair.

"I have to intervene now or we won't sleep a wink tonight," I say.

"Don't do it," Joseph says. "We will get beaten up."

Fortunately, I worked as a bartender for six years during college, so I know exactly how to deal with these types of people. I approach the group and, after complimenting them on their choice of music, I introduce myself.

As expected, the word *Amsterdam* works miracles.

"*Amsterdam party city, legalize marijuana!*" shouts one of the heavily intoxicated men.

Instantly, I am their best friend. That was step one: Every bartender knows that you have to make friends with a drunk first, if you want to make him leave of his own volition. The aggressive approach, in which you grab him by the neck and hurl him into the street, can have disastrous consequences. Drunks don't give up easily.

Not long after that, I am the center of the Kenyans' attention. They all want to take a selfie with "the boy from Amsterdam." I willingly go around the group and pose for selfies, trying very hard not to get groped by the ladies. My traveling companions are watching the whole scene with suspicion; they seriously doubt as to whether I'm still on their side. But then I proceed to step two.

"Hey guys, we are having a great time, but we really have to go to bed now. We have to drive all the way back to Nairobi early in the morning to celebrate his mother's birthday." I point to Ethan.

Every bartender knows that with the mother card you can play on a drunkard's feelings; they generally are extremely sensitive to that.

There is a murmur of agreement, and after everyone has had their photo taken with their new best friend from Amsterdam one last time, the partying group leaves for the other end of the field, after which peace returns to our encampment.

When we are in our tent we hear only the bass of the music in the distance. Our tent is yanked a few times to ask if I can come and join their party.

"He's sleeping," I answer in a twisted voice. "He will see you tomorrow."

Maasai Mara, Kenya

The Maasai Mara is a vast grassy plain of almost 2,000 square miles in southwestern Kenya. It is bordered by the nearly nine-times-larger Tanzanian Serengeti in the south. The areas are separated by the Kenya-Tanzania border, but they belong to the same ecosystem.

This endless plain is the setting for the "great migration," one of the most impressive natural spectacles in the world. Nearly 2 million wildebeest, 400,000 Thomson's gazelles, and 300,000 zebras follow the annual rain cycle in a giant circle through this vast area; the phenomenon is known as the Circle of Life.

In January and February, the herd, coming from the Ngorongoro plains, arrives in the southeastern part of the Serengeti, where it has just rained and the grass is greenest. More than 400,000 calves are born in a very short period of time. Until the end of March, the young animals have time to grow stronger for the journey of a lifetime that from the moment of departure will never end. The lives of the newborn animals are precarious; the tall grasses are an excellent hiding place for hungry lions, cheetahs, leopards, and hyenas.

At the end of March, they are forced to leave, as by then all the grass has been eaten and the ground is completely bare. The calves that have survived their first months of life follow the rest of the herd toward the Seronera Valley, in the central part of the Serengeti. From here they set course to the north. The Circle of Life is rejuvenated because here the adult wildebeest, zebra, and gazelle females are mated by the males. On the way there, they cross the Mbalageti and Grumeti Rivers, where hundreds of animals disappear between the jaws of Nile crocodiles who are a few dozen yards long.

After a long journey, the herd arrives in the northern part of the Serengeti around the end of July. A few hundred thousand animals remain here, but a large part of the herd continues further north. The next obstacle is the Mara River. Driven by instinct, thousands of animals jump one after the other from the tall riverbank into the swirling water, several yards below. Some animals break their legs and are carried away by the current, others are torn apart by crocodiles. The animals that make it across are met by hungry lions and hyenas as well as by camera crews from all over the world.

At the beginning of August, the herd finally arrives at the Maasai Mara, where the rains have just fallen. Until the end of September, the animals can feast on endless plains of juicy green grass. When the plains are eaten bare, the herd leaves for the south again. A long and arduous journey, especially for the animals that are heavily pregnant at the time. Via the Ngorongoro plains, they migrate in a wide arc all the way to the southeastern part of Serengeti, where the Circle of Life begins again in January and February.

Normally, you pay hundreds of euros a night to stay at the Encounter Mara Camp, but I made a good deal with them: We can stay there for a few dozen euros in exchange for promotion on my blog.

The entire camp is run by traditional Maasai, the original inhabitants after whom the Maasai Mara is named. When Dixon, a man dressed in scarlet robes, with red ocher in his hair, giant rings through his ears, and a spear in his hand, escorts me to my tent, I compliment him on his beautiful outfit. He smiles and says that the Maasai are a proud people who place great value on traditions.

My tent can be called a tent only because the roof and the walls are made from canvas, but it is more of a luxury apartment. On either side of the bed are handmade wooden statues of Maasai warriors, and the bathroom has an open connection to the outside, giving you the feeling that you are in the middle of the African bush. And in fact you are. When I hear an elephant trumpeting in the distance while taking a shower, I feel like Robert Redford in *Out of Africa*.

During dinner, we are joined by Dixon and three other Maasai. We talk about their culture and their perception of the Maasai Mara as a game reserve.

The vast majority of tourists book all-inclusive tours from Nairobi. They are transported by safari vans or private planes to the Mara, where they start their safari from a luxury lodge. The only Maasai they see up close sell necklaces and figurines at tourist shops that are disguised as traditional Maasai homes. The vast majority of revenue ends up in the pockets of foreigners or corrupt government officials.

Local residents benefit far too little from the gigantic flow of tourists, which gives rise to conflicts. Many Maasai simply do not see the point of the Mara as a national park, because they are not allowed to graze their cattle in a large part of the area.

There is no fence around the Maasai Mara—which is great, as it allows the millions of ungulates on the Mara to migrate freely, but it also means that people can walk into the national park with their livestock, even if this is only limitedly allowed. For a pride of lions, such a herd is like a walking buffet. The Maasai see their cows as their most important possessions, so when one is killed by lions, it's a real drama. If this happens within the boundaries of the national park, they are not entitled to compensation from the government and, therefore, some Maasai decide to take the law into their own hands. They chase off the lions from the carcass of a cow they've killed and poison the carcass, causing the lions, which inevitably return to continue their meal, to die gruesome deaths. And it's not just lions that have suffered: A poisoned cow carcass was recently found surrounded by dead jackals and hyenas and by more than a hundred dead vultures from five different species, most of which are critically endangered. This is one of the main reasons why these birds are declining so rapidly in East Africa.

The next morning, we drive up the Maasai Mara in an open Land Cruiser. Dixon, who knows the area like no other, is behind the wheel. It is April and, therefore, the main display of the great migration is currently unfolding in the Serengeti. The large herds of wildebeest are missing, but it is nevertheless a great experience to drive through such a vast grassy landscape. At some points we can see dozens of miles in the distance. Only lone acacia trees stand out here and there on the horizon; otherwise the relief in this vast landscape is formed by herds of antelopes, zebras, buffalos, and elephants.

When I fly to Uganda at the end of April, my count stands at 2,851 bird species. Kenya has been extremely prolific, but Africa has many generalists—bird species that have a wide geographic distribution and are found in many different types of habitat. If you travel through Africa over a longer period of time, you will often encounter the same species over and over again, and it becomes increasingly difficult to come across new species.

While I'm waiting to board the plane, I browse through *Birds of Africa South of the Sahara* and come to the conclusion that I have already checked many species. I am haunted by the feeling I have allocated too much time for this continent. Time that I could have spent much better somewhere in Asia or Australia. But my flights are booked and paid for, and there is absolutely nothing I can do about it.

Entebbe, Uganda

I paddle in a wobbly canoe through the Mabamba Swamp, along the shore of Lake Victoria, accompanied by four birders and a local guide. The morning mist still hangs over the papyrus reeds. A few hours ago I landed in Entebbe, and once again I got barely any sleep. Yet excitement supplants my fatigue; somewhere in this vast swamp, I hope to encounter the Shoebill.

The fact that birds are direct descendants of dinosaurs can be clearly seen from the Shoebill. The Latin name for this bird is *Balaeniceps rex*, like the *Tyrannosaurus rex*. *Rex* is Latin for king, and while the Shoebill may lack the razor-sharp teeth of the T-Rex, that doesn't alter the fact that he is the undisputed king of the papyrus swamps of Central Africa.

This solitary, stork-like bird has—as the name suggests—a gigantic, clog-shaped beak with which it hunts all kinds of fish. Sometimes it also likes to go for unheeding waterfowl. It is even rumored that an antelope calf once disappeared into a Shoebill's stomach. Its enormous size, grotesque beak, light irises, and blue-gray plumage give it a prehistoric appearance.

A single pair of Shoebills covers a vast territory. While nesting, they are extremely sensitive to disturbance, and it is, therefore, not surprising that they are suffering greatly from Uganda's explosive population growth. On average, a Ugandan family consists of five or six children. To feed this ever-growing population, the papyrus swamps are being drained and cleared for agriculture. In short, the Shoebill's habitat is disappearing. Moreover, pollution, as a result of oil extraction, and capture for the illegal cage-bird trade are having increasing impacts.

No one dares to speak as we sail across the narrow creek through the papyrus swamp. The Shoebill is a shy bird and will take to its wings at the slightest disturbance. After every bend, we peer into the vegetation, hoping that we will suddenly come face to face with the king of the swamp. We see different types of herons and kingfishers, and we regularly encounter an African Jacana, a brown-and-white, rail-like bird with a light blue beak and extremely long toes, which allow it to walk effortlessly over floating aquatic plants. The Shoebill, however, despite its size, seems to be untraceable.

After more than two hours of paddling, we are lucky. Rounding the umpteenth bend, a gray figure suddenly appears, standing exposed at the edge of the vegetation about 20 yards away. It is a male, recognizable by its gigantic beak—a female Shoebill has a noticeably smaller beak. He doesn't move at all. The fluttering tuft on the back of his head gives him a stately but, at the same time, clownish appearance. He is clearly watching us, but as we drift along with the current in silence, he doesn't feel threatened, giving us the opportunity to examine him in detail. This particular Shoebill has managed to find a partner in the swamp, because a little later we discover a female, which is a bit more hidden between the papyri. This is a good sign; Shoebill couples stay together for a lifetime.

⁓–⁓

Murchison Falls National Park is located in northwestern Uganda, at the border with the Congo River, about half a day's drive away. On our way, there isn't one dull moment, because the savannah landscape is really bursting with birds. Stately groups of Grey Crowned Cranes regularly line the road, and we see Marabou Storks in almost every acacia tree. With their bald, warty head and pink throat pouch, these stork-like creatures are arguably the ugliest of all birds.

After a few hours driving, our van starts to act up. The temperature gauge on the dashboard moves way too far to the right. Ibrahim, our driver and guide, stops the van and opens the hood. He looks concerned.

"The fan is broken and the coolant has run out."

You don't want to hear this when you're stranded in a remote part of Uganda. A car with such defects wouldn't be allowed on the road anywhere else, but this is Africa, and we reluctantly hand over our drinking bottles to fill the cooling system with water.

This plan seems to work for a while, and we continue at a normal pace, until the temperature gauge creeps slowly to the right again and the driver has to park the van at the side of the road for the second time. This time, a thick column of steam shoots high into the air as he unscrews the cap of the coolant reservoir. The engine is now red hot, and it takes us more than half an hour to refill the reservoir. The rest of the day we can drive for only half an hour at a time, after which we have to stop and wait for half an hour.

When we arrive at our destination just before midnight, there are no garages in sight. We fill two large jerry cans with tap water as a precaution, so that we won't have to sacrifice our precious drinking water the next day.

⁓–⁓

We often see large family groups of baboons along the road. While they're busy grooming, they give us uninterested looks with their reddish-brown, beady eyes. Some females carry helpless-looking young on their backs. The dominant males, nearly twice the size of the females, have broad shoulders and dangerous-looking jaws. Every now and then, one such specimen sits in the middle of the road, like a Buddha statue, his pink willy demonstratively on show. I chuckle, as the image reminds me a bit of Gandalf in *The Lord of the Rings*. "You shall not pass!" He starts to move only when we are a few yards away. He then reluctantly gets up and walks painfully slowly to the side, where he shows us his contempt by presenting his ass.

Gradually, the savanna gives way to tropical rainforest. This transitional area is the habitat of a shy, brown-colored bird without obvious field marks, which is especially notable for its loud, whistling song: the Puvel's Illadopsis. A real birders' bird—normal people wouldn't even bother to pick up their binoculars for such a "drab," nondescript species. In the Netherlands, we have lots of these types of birds, such as pipits, larks, reed warblers, and buntings. They are all brown and inconspicuous. But look at such a bird through binoculars or a telescope, and you will see beautiful, subtle details in its plumage. They are much more beautiful than, say, a gaudy parrot, if you ask me.

At sunrise, we hear one singing. One of my traveling companions takes out his smartphone and briefly plays the song of the illadopsis: "*Tjieuu-Tjiet-Tjieutjiet-Tjiutjitjutjutju!*"

The call worked, and after a few minutes the song of the illadopsis suddenly sounds a lot closer. We stare tensely into the thick bushes, but then I feel a fierce bite on my neck and then on my arm. A loud "Ow!" from my companion makes it clear that he, too, has been bitten. A huge swarm of tsetse flies buzzes around us. I absolutely hate stinging flies; soon I forget about the illadopsis, slapping around me like a madman. We try to hold on for a while, but the incessant attacks are driving us mad and we run back to the lodge. One person remains stoically in position.

"Don't be such babies. It's only flies, and a few bites have never killed a person." What he apparently doesn't know is that a small percentage of tsetse flies carry the trypanosomiasis parasite, which can cause the deadly sleeping sickness.

On our way north, we hear a familiar sound right alongside the road. We stop and pick up the smartphone again. But as soon as we open the windows, dozens of bloodthirsty stinging flies zoom in. They immediately attack our bare arms and legs. We go ahead very quickly, and within a few seconds a loud *"Tjieuu-Tjiet-Tjieutjiet-Tjiutjitjutjutju!"* reverberates from the phone's speaker. Almost immediately a Puvel's Illadopsis lands on a branch nearby, where it remains for a short while. Luckily it doesn't linger much longer, as by now, the van is invaded by biting insects.

We drive on quickly as we try to knock the flies away. Just when we've swatted most of them and think we've finally got rid of them, our van slowly comes to a halt, and we're overtaken by the swarm again. I forgot for a moment that the engine cooling system is still not working.

Just like the day before, in the hours that follow, after driving for half an hour, we have to take a break of half an hour in order to let the engine cool down. In the meantime, we wage a continuous, unwinnable battle with a tireless army of tsetse flies. This safari will go down in the books as the most uncomfortable of our lives.

When we arrive at our campsite in the northern section of the park in the afternoon, we have donated at least 40 ounces of blood to the local stinging-fly community. Thank goodness the landscape here is a lot drier, which means that the biting-insect swarm has disappeared.

In the evening, we take a trip on the White Nile. The experience of this oasis of tranquility on the water is in stark contrast to this morning's hellish ordeal. Along the banks, we see buffalos and elephants and unimaginable numbers of waterfowl, and to our great surprise we encounter another Shoebill.

The boat trip ends at the world-famous Murchison Falls, which is almost 40 meters tall. Every second, almost 80,000 gallons of water pour down through an opening of only a few yards. We float at the foot of the waterfall, and from so nearby, the fierceness of nature is extremely impressive. All of a sudden, two graceful, tern-like birds land on the smooth rocks next to us. They are Rock Pratincoles, beautiful waders with black masks, white stripes on the cheeks, and bright red beaks. They immediately have the full attention of our group; nature's ferocity no longer matters. This is the beauty of birding:

When a special bird is sighted, everything else around you fades into the background, except for stinging flies.

In Uganda, things are not going the way I would like them to. I'm missing two of my most desired species, the Green-breasted Pitta and the highly localized Grauer's Broadbill. Car troubles, stinging fly infestations, and failing guides have not helped. The daily number of new species is quite stagnant. My suspicion of ten days earlier seems to have come true: I have allocated far too much time to East Africa. My next destination is Tanzania, via Kenya. When I go through my bird book, I come to the conclusion that, if everything goes really well, I will see at most seventy new species there.

Namanga, Kenya

My phone is playing the song "I Like Birds" by Eels. Stratton is behind the wheel of his old Land Rover and I slouch next to him, my bare feet dangling out the window, the warm African air blowing over my legs. Joseph, who after the Kenya adventure decided to join this trip as well, sits in the back of the 4×4.

"This old banger has been through a lot," Stratton says with a laugh. "An aggressive male elephant once demolished its entire rear end during a safari in Tsavo National Park. I saw that beast's tusk pierce the left back door. It was just two feet away from my head."

Stratton is a twenty-five-year-old American who has lived in Africa for most of his life. First in Angola, during the civil war, then in Zimbabwe, and finally in Kenya, where his parents are involved in development cooperation. He spends most of his time in the Maasai Mara, researching Martial Eagles. He is fluent in Swahili, always walks around in sandals and khaki shorts, and seems more African than American.

I met Stratton about two years ago through Birding Pal, just like Ethan. While visiting his Dutch girlfriend, Stratton was looking for someone to go birding with. We spent a cold winter day exploring the Rhine delta, and we've been good friends ever since. Whenever he is in the Netherlands, we try to make time for a day of birding.

And now, we are traveling to Tanzania together.

"Man, you can't imagine. I was going crazy behind my computer. The prospect of this trip was the only thing that kept me going in this exam period. This is going to be such a cool adventure!"

His infectious enthusiasm boosts my confidence again.

As we are driving the last hundred yards up to the border, we hear a loud bang against the side of the car, followed by the hysterical screaming of a woman in the street. From all sides, bystanders run toward our car. From the rearview mirror I see an old man lying on the street with his bicycle. This is looking bad—it seems we hit someone.

If you run over someone in Africa, it can lead to a dangerous and potentially fatal situation. An acquaintance of my father experienced this personally during a working visit to Sudan. As they were leaving a village, the driver that followed them hit a small child. The occupants got out to help the child but were immediately dragged away by an angry mob and stoned to death on the spot.

Fortunately, I see from the rearview mirror that the old man has scrambled to his feet. Meanwhile, Stratton and Joseph get out to help him, while calming down the bystanders. "We will take this man to a hospital and pay him compensation."

The word *compensation* has the desired effect, and five minutes later the victim sits in our car, bicycle and all, as we drive to the nearby field hospital, escorted by a group of bystanders.

The old man gets out and limps to the entrance, leaning on Stratton's shoulder. After a short while, they come out again and shake hands.

"It's settled," Stratton says, getting back in. "I gave him a thousand shillings [about ten dollars] to cover his medical expenses. Let's move fast before we get into more shit."

I see the old man limping back to his bike, but halfway there, he seems to have magically recovered and he does a happy dance. He forgets to do a wheelie as he cycles away.

"A road trip in Africa isn't complete without some good hustling bwana," Stratton says with a laugh as we drive into Tanzania.

From the lowlands, we hit a very bad road that meanders up. It is littered with deep mud puddles, huge boulders, and broken tree branches. Stratton is completely in his element. "A Land Rover is a bit like a dog; you have to let it roll in the mud every now and then, otherwise it will get old and grumpy."

On arrival at the camp, we are welcomed by Arthur, the owner. With his Rhodesian accent, beige tropical hat, and long gray mustache, he is reminiscent of Van Pelt, the murderous hunter from *Jumanji*. Stratton and I both get a handshake, but Joseph does not. I get the bad feeling that this has something to do with Joseph's skin color.

During lunch, we're joined by Arthur. He likes to have an audience and talks with great aplomb about how, years ago, he and his wife ran a luxurious hunting lodge in Rhodesia, present-day Zimbabwe. Due to reforms in that country, they were forced to close it.

"We had dozens of servants and used to organize hunting expeditions for millionaires. Those trips cost thousands a day, and we hunted everything from elephants to lions."

Dirty bastard, I think to myself.

Arthur continues, "Would you like to meet Dr. Livingstone?"

I look at him questioningly.

Arthur shakes his head with a laugh. "This Dr. Livingstone is a bush baby, a kind of nocturnal prosimian. We are the only ones who have ever managed to raise such an animal as a pet."

Stratton nudges me when Arthur disappears into the kitchen. "I have several friends who have bush babies as pets, so I would take his stories with a grain of salt."

"You both can come," Arthur says to Stratton and me when he returns. "But unfortunately, he can't come in with you."

Arthur points to Joseph.

"Dr. Livingstone can't stand people with a black skin."

Did I really hear this? I am dumbfounded. I imagine seeing a bush baby in a KKK robe.

I feel uncomfortable when I walk behind Arthur into the house. I immediately see the bush baby sitting on top of a display cabinet. For a few seconds, he gives me a suspicious look with his saucer-shaped, brown-orange eyes, but then he suddenly makes a wide jump onto a lamp to end up on my shoulder.

"Hello, little Dr. Livingstone," I say to the bush baby. "You're a sweetie, aren't you?"

Almost immediately, I feel Dr. Livingstone's sharp teeth in my neck. I try to pull him off, but he holds on tight and bites until I'm almost bleeding.

"Dr. Livingstone, get off!"

Arthur pulls the eager prosimian from my shoulder, and I escape unscathed.

At least, the bush baby makes no distinction in skin color, in contrast to his owner.

Our days in the Usambara Mountains turn out to be long and hard. There are hardly any paths, so we are forced to plod through the forest, climbing up steep slopes. I see only a few new species in three days. However, we do see some rarities, including the Usambara Eagle-Owl. After a nightly search, we hear a male in the dense tree-tops, high above our heads, calling for a female. The sound—a hollow, accelerating *"WHÒ… Whò… Whò… Whò-whò-whò-whò-whò-whò-whò!"*—rolls like thunder through the night.

It takes at least another hour before we finally spot the bird, but we see it completely exposed, perched on a thick, gnarled branch: a giant owl with a black-spotted chest and an orange face mask and dark eyes that glow bright red in the beam from my flashlight.

During our last morning in the Usambaras, I can't seem to lay my eyes on any new species, and when, after a long drive, we arrive at Arusha Airport in the late afternoon, my count still stands at zero. Therefore, May 16 will go down in the books as the first official *zero day*. My lead over Noah is slowly shrinking, and I start to get worried. How on earth am I going to see enough new species in Africa over the next month and a half?

Antananarivo, Madagascar

Madagascar, sometimes called the eighth continent, is one of the largest islands in the world. It broke away from India about 88 million years ago. Because it has been geographically isolated for such a long time, the plants and animals there have developed in a unique way. Ninety percent of its animal life and 80 percent of the plant species exclusively occur there. Most notable are the lemurs, also known as makis, of which more than a hundred different species are known, and new species are discovered almost every year. These prosimians show a great variety in shape and appearance: Some are as small as a mouse, and others are as large as a baboon. But one thing they have in common is that almost all of them are endangered. At least seventeen species have become extinct since humans set foot ashore 2,000 years ago.

It should come as no surprise that the birdlife in Madagascar is equally unique. Six of the world's more than 240 bird families can be found only in Madagascar, and more than 60 percent of all species on the island do not occur anywhere else.

As a little boy, I watched a number of BBC nature documentaries in which David Attenborough introduced me to lemurs, vangas, and ground rollers, and I became fascinated by this magical island.

I land in Antananarivo in the middle of the night. Immediately upon arrival, I am confronted with the deep-rooted corruption. A customs officer demands that I pay cash for my visa, and I receive a pile of dirty, discolored banknotes in exchange. When I later want to buy something in a roadside restaurant, it turns out to be counterfeit money. Being ripped off by a customs officer at an international airport is the pinnacle of corruption. Then again, Madagascar is one of the poorest countries in the world; more than 90 percent of its population live on less than two dollars a day.

"Make yourself comfortable. It's a nine-hour drive to Ankarafantsika National Park," my driver says, throwing my backpack in the trunk.

As it will still be dark during the next few hours, I try to catch up on some sleep in the back seat of the Land Cruiser. However, it is a winding road, and at every bend, I slide from left to right, falling half on the floor. After sunrise, sleeping is no longer an option anyway; now that I'm finally here, on this magical island, I don't want to miss a thing.

I had thought Madagascar was mostly forest, but all I find are grassy hills as far as the eye can see. Nothing but grass, farmland, and alien plantations. The fact that nature has been destroyed at such a scale is almost inconceivable. It sometimes looks like a moonscape. Gullies of red soil cut through the bare ground. Tree roots used to hold the earth together, but now devastating mudslides caused by heavy rainfall leave gullies several yards deep scarring the landscape.

———————

We finally arrive at our destination. I yawn as I get out of the car. My legs are stiff and my back hurts from the long drive. You have to be prepared to pay the price, I think to myself. Suddenly, I see something moving in the trees around the parking lot. I quickly grab my binoculars and can barely believe my eyes: a Coquerel's crowned sifaka! Rarely have I seen such a beautiful animal. These sturdy lemurs have long tails and are covered from head to toe in woolly, snow-white fur, which stands out sharply against their black faces and chestnut-brown thighs, chest, and upper arms. It is a family of about ten animals. Among them is a cub—a fluffy, helpless moppet that clings to its mother's back.

As I'm watching the lemurs, out of the corner of my eye I see a flock of black-and-white birds landing in a treetop. Sickle-billed Vangas, one of the species I am dying to see. I feel like a kid in a candy store, even though I haven't even set foot in the forest yet!

———————

Ankarafantsika literally means "mountain of thorns." An apt name because the bone-dry forest around us consists largely of trees and shrubs with razor-sharp thorns. My guide, an elderly man from the village nearby, walks silently ahead of me along a narrow, sandy forest path. His stride is slow and calculated; he is constantly looking around and attentive to every sound. You can see from the way he moves that he has been walking in this forest all his life.

Then there is a loud, nagging call, resembling the sound of a bird of prey: "*Kiuuw-Kiuuw-Kiuuw!*" The guide freezes and points forward: "*A Cuckoo-roller!*"

I'm on edge. This bird is so unique that scientists have placed it in a monotypic family—a bird family consisting of only one species. I sneak up behind him in the direction of the sound, which gets louder with each step we take. Then I see the Cuckoo-roller; he's not even 5 yards away. I've never seen a bird even remotely resembling this one: a sturdy, light gray animal with an enormous head and a glossy green back. His posture is somewhere in between a European Roller and a hawk, making him resemble a bird of prey. A few yards farther, we see his brown-spotted female. Cuckoo-rollers are not shy, and they play an important role in Malagasy mythology. They are seen as messengers of love, as they are almost always seen in pairs.

I give my guide the thumbs-up, but he's watching the birds, completely mesmerized. He has probably seen a Cuckoo-roller hundreds, if not thousands, of times, but he obviously doesn't mind. That's what characterizes a good guide: He's just as happy with a sighting as I am.

In the evening, we make a boat trip on an idyllic lake near where we are staying. Thousands of lilies float on the smooth surface of the water, and we see night herons and Squacco Herons in the reeds along the banks. This is one of the last places in the world where the Madagascar Fish Eagle is reported to breed. We see this enormous brown-gray bird of prey at a distance of less than 20 yards, eating a fish in an old, overhanging fig tree. It clutches its prey to a branch with its claws and tears it to pieces with its dangerous-looking hooked beak. I couldn't have imagined a better finish to an already great day.

At nightfall I treat myself and the guide to a cold beer and take stock. Today, I have seen five new bird families, spread over thirty-eight different species. After those unproductive days in Tanzania, I really needed this boost.

Andasibe-Mantadia National Park, Moramanga, Madagascar

The bumpy drive through Madagascar, which lasted nine hours, was a breeze compared to the journey to Andasibe-Mantadia, which takes fourteen hours and driving through the night.

When I get out of the car and stretch my cramped muscles, I feel the cold wind on my face; the contrast with the dry heat of Ankarafantsika could not be starker.

According to a park ranger at the entrance, the temperature here has dropped more than ten degrees in recent days. This is a setback. I had prepared for it to be wet, but the strong wind and cold make me feel like I'm walking around the eastern tip of Vlieland at the end of October: not exactly the best conditions to look for the Helmet Vanga, the holy grail of Madagascar.

In the back seat of the 4×4, I'm shaken about incessantly. It feels like I've been stuffed in a washing machine. We left this morning at three o'clock, and it hasn't stopped raining since. The headlights shine on the most miserable road I've ever seen. "Road" is an overstatement, as it looks more like a boulder-strewn mudslide. The speedometer shows only 6 miles an hour, and we have still more than 18 miles to drive to the site of the Helmet Vanga.

The plan was to start our search at dawn, when the vangas are most active, but we only arrive at the site around 7 a.m. When we get out of the car and get ready to walk, I notice that I have forgotten both my packed lunch and my water bottle: It's a huge 2–0 disadvantage before we have even started.

We walk into the forest on a narrow path through the pouring rain. A man from the village, armed with a machete, takes the lead and is closely followed by the guide and myself. After a few hundred meters, we have to leave the trail and cut our way through the dense vegetation and move up along a steep slope. I am soon covered in mud and completely soaked with sweat and rain. A few times we think we hear a Helmet Vanga, but every time it turns out to be a Crested Drongo, which flawlessly emulates its song.

We continue like this for hours. Slope up, slope down, straight through the dense, soaking wet rain forest. Because I forgot my lunch and water, I have to keep myself going with chewing gum and I am forced to lick raindrops off leaves. We hardly see any birds, at best one every hour, and by midafternoon I am completely wrecked. I ask how far it is to walk to the car.

"Three miles," is the answer.

Three miles. Straight through the densest, wettest rainforest imaginable ...

Finally, we reach the 4×4 at dusk without sighting the Helmet Vanga, defeated by the forest. Up to this point, I have managed to get through all the hardships. But this miss, combined with my chronic fatigue, sore body, and soaking wet clothes, tests my perseverance to the limit. Hours later, when we arrive back at the lodge, I call Camilla and tell her that I feel completely shattered, that I miss her, and that I can't wait to see her in South Africa. Suddenly, the two weeks that still separate us seem to be unbridgeable.

The next morning, I have two hours left in Andasibe-Mantadia before I have to return to Antananarivo to catch my plane. The rain is still pouring down and I manage to see only a few new species. I'm at the end of my tether, all my stuff is soaking wet, I don't have a single piece of clean clothing left, I have a cold, and I miss my own warm bed.

I thank the guide for all his hard work and step into the car, soaked from the rain. Along the way, I try to dry my clothes by hanging them over the chair backs, and stuff myself with vitamin preparations, Tylenol, and Antigrippine. With my cutthroat travel schedule, I can't afford to get sick.

A little later I arrive in Antananarivo, from where I will fly to Nairobi, Kenya, where I will have to wait for three hours, before continuing via Lusaka, Zambia, to Lilongwe, the capital of Malawi.

Once on the plane, it all gets a bit too much for me. The bad weather and fatigue have dented my confidence. I'm starting to doubt the whole venture and wonder if I'm going to make it to the end of the year. My thoughts race through my head and sweat beads up my forehead. I feel like I'm having a panic attack. I put on my earplugs and take a Xanax to calm down. Thankfully, there's Bob Marley's raspy voice: *Don't worry about a thing / 'Cause every little thing gonna be alright.*

Lilongwe, Malawi

The arrivals hall is lit by only a flickering fluorescent tube, and when I go to the toilet there is no running water to wash my hands. I have to wait at least an hour for my visa to be stamped and signed by a customs officer. It is a yellowed piece of paper with some illegible scribbles on it, crookedly stuck in my passport.

I take a taxi to a hotel at the edge of town. I have to ring the doorbell for half an hour before a sleepy night porter finally opens the door. I stagger to my room and fall asleep like a log. It's three o'clock in the morning.

When I turn off my alarm, I see that I slept for only forty-five minutes, a record this year. So I don't feel my best, to put it mildly.

In the breakfast room I am greeted by Rogier, my travel companion for the coming week. From the very beginning, he has been a great support in the realization of my plans. With his law firm, he decided—as one of the first—to sponsor me. He often helped me brainstorm about how I could get more attention for my charity, and he even organized a sponsor evening at his office. He is also an avid birder and a great aficionado of Africa, so he decided to visit me in Malawi.

With deep bags under my eyes, I have started on a dry croissant and I'm on my third cup of strong coffee. Today, things just don't seem to be working out. Rogier, on the other hand, is in a great mood. At home he has a busy job and three children, so he rarely gets the opportunity to go birding.

"I'm enjoying it already. I've been looking forward to this for months."

His enthusiasm is contagious, and I'm actually starting to feel a little better. The importance of new travel companions cannot be underestimated. Time and again, their enthusiasm gives me new energy and the strength to go for it again. Moreover, it is great if you can share experiences and sightings with someone else.

Dzalanyama Forest Reserve is known for its miombo forest, a forest of low deciduous trees and ground vegetation of knee-high grass. It is found only in south-central Africa, from Angola in the west to Mozambique in the east. Few people have ever heard of this miombo forest, but it is very well known among avid birders: It offers habitat to more than forty bird species that cannot be found anywhere else.

The miombo forest plays an important role in the daily life of small communities in rural Zambia and Malawi. The trees produce fruit, and the dry grass that grows between the trees is used as roofing material and animal feed. Unfortunately, it is precisely these communities that have suffered the most from the ongoing heat wave caused by El Niño, which led to crop failures and forced people to find other ways to make ends meet.

When we drive into Dzalanyama, we regularly see people walking out of the forest with their bicycles, carrying logs on their luggage racks, sometimes piled several yards high. They burn the logs into charcoal, which they then sell along the road and at the market in Lilongwe.

The lodge is located in the middle of the forest. It is a simple wooden house, with a kitchen where we have to prepare our own food. I love it; luxury is unimportant to me, as long as I am surrounded by nature, in the best birding spots. I plop down on a couch and close my eyes for fifteen minutes. These kinds of power naps are indispensable to get through a day like this—after such a hard night of travel—otherwise a breakdown will sooner or later throw a wrench in the works.

It is the middle of the day and hence very hot and windy as we walk into the forest. During the first hour, we see nothing at all, although there should be so many good species here. The effect of the power nap starts to wear off, and soon I am trudging through the blazing sun, feeling completely drained. Fortunately, at the end of the afternoon the wind dies down and the sun sets behind a mountain. Suddenly the forest is cool and shady and this change is noticeable in the bird activity: We start to hear some hesitant birdsong left and right. Slowly the forest comes to life.

Two Fork-tailed Drongos fly across the path ahead of us, followed by another, and just after that a whole group. There is a range of different bird sounds coming out of the forest. I think back to one of the first days of this year, in the Sinharaja Forest Reserve in Sri Lanka, when we ran into that gigantic mixed species flock.

I mimic the call of a Pearl-spotted Owlet: "*Tu tu tu tu tu tu.*" I learned this trick from Ethan. This little owl hunts passerines, which see it as a fearsome enemy. If the passerines see or hear one of these owls calling, they try to chase the owl away as quickly as possible. Because the owl hunts from cover, once he's discovered, his chances of catching a meal are gone.

My trick works because birds come flying in from all directions. Soon the trees around us are full of alarmed passerines. One of the first species we get a good look at is the Anchieta's Sunbird, a 4-inch bird with a downward-curved bill, a blue face, and an unmistakable red-and-yellow belly. A bit farther, an Orange-winged Pytilia is scurrying across the ground, and above us sits a male Broad-tailed Paradise Whydah. The latter is a species-specific brood parasite that lays its eggs exclusively in the nest of the pytilia. The eggs of the whydah and the pytilia are hardly distinguishable one from another and are incubated by the unsuspecting pytilia, which then instinctively takes on parental care.

It's all going very fast. When the flock has disappeared into the forest again, I grab the bird book. It turns out that, in less than half an hour, we've seen thirty-four different species, including many new ones.

It's getting dark, so we return to the lodge, where we open a beer and toast to the success of this afternoon. I take off my shoes and rest my tired feet on the railing of the porch. What a wonderful moment after the chaos of the past few days. As I take a sip and blissfully stare ahead of me, I hear the call of an African Barred Owlet, the thirtieth new species today. My confidence is back, and I even get a bit of a holiday feeling.

Liwonde, Malawi (1995)

When I was in the sixth grade, my parents went to Malawi for two weeks. There they visited the Kafulafula Primary School which consisted of only three dilapidated mud buildings with no lighting, where more than a thousand children from the surrounding villages attended school every day. They had to sit on the floor, as there was no money for desks. This experience left a deep impression on my parents.

At that time, my mother was on the parents' board of my primary school. During a meeting, she suggested organizing a fundraising campaign for the Kafulafula Primary School at the yearly summer party. It was a great success: We raised more than 16,000 guilders—the equivalent of $7,500—in total.

Six months later, my parents and I left for Malawi with a load of goods weighing almost 450 pounds. These materials included typewriters, clothes, volleyball nets, pens, notepads, blackboards, and inflatable footballs. When we arrived at the school, the construction of three new classrooms had already started. My father was a structural engineer and had designed those rooms in such a way that the 16,000 guilders would be just enough to finance the construction project.

News of our visit spread like wildfire, and that same day we were invited by the nearby Mvuu Camp to spend a few days with them in Liwonde National Park.

The camp was located along the Shire, Malawi's largest river. The banks were truly teeming with birds. I saw bright blue Malachite Kingfishers chasing fish from overhanging reeds and colorful bee-eaters flying overhead by the hundreds. In the evening, we went on a night drive; we drove around in a 4×4 equipped with a big spotlight, looking for nocturnal animals. We saw different kinds of owls and nightjars and even a leopard, which walked ahead of us for several minutes in the glow of the headlights.

Liwonde, Malawi (2016)

From Dzalanyama, it is about half a day's drive to Liwonde National Park. We pass villages consisting of mud houses with thatched roofs. I see children playing a game of soccer on a dusty football pitch. A bit farther down, a man is working his field with a wooden plow, which is pulled by a cow whose ribs are clearly visible.

It strikes me how dry everything is. Lakes that once burst with life have turned into barren voids of cracked, bone-dry clay. Vegetable gardens are dead and abandoned, and all around we see mango and papaya trees with withered yellow leaves. It is clear that this country is suffering from the severe drought.

On our way there, we have a short stop in Lilongwe to exchange money. This takes place in a parking lot at a supermarket, not exactly the border exchange office I had in mind.

We give our guide three hundred-dollar bills. He gets out and returns fifteen minutes later with a shopping bag full of paper money.

"I think we just became millionaires," Rogier jokes.

We have to laugh at our newly acquired millionaire status, but this inflation is of course worrying. Malawi is the sixth poorest country in the world, and more than 65 percent of the population lives below the poverty line (by the way, the five even poorer countries are all in Africa, namely Niger, Liberia, Burundi, DR Congo, and the Central African Republic). The ongoing drought makes the situation even more desperate, as a third of their gross domestic product comes from agriculture. If the drought continues, and it very much looks like it will, more than 6.5 million Malawians will soon be dependent on humanitarian aid.

Despite this crisis, the people here are remarkably friendly and hospitable. In the West, we can learn something from that. Children standing along the road wave at us, and when we stop for something to eat or drink, people come up from all sides to welcome us. A cynic would say they do that to get money from us, but that's definitely not true. When I explain that we are here to watch birds and show my bird book, people immediately start to flip through the book and enthusiastically point out species they have seen in the area.

A little later, we drive by the Kafulafula Primary School. The school buildings, designed by my father twenty years ago, are still there. Dozens of laughing children in neatly washed and ironed school uniforms run around the schoolyard. My thoughts go to my parents; thanks to them, these children receive education with a decent roof over their heads.

Mvuu Camp has remained almost exactly as I remember it. The canvas tents overlook the river, where herds of hippos swim closely together and Nile crocodiles, several yards long, lay completely still on the banks with their mouths half open. In the afternoon, groups of African Skimmers fly low over the water, heading for their roost. I feel like we're in paradise.

That evening, we go on a night drive. Our target is the Pel's Fishing Owl, one of the largest owls in Africa. During the day they sleep along the river, well hidden under the dense canopy of old fig trees, and they emerge at night to hunt frogs and fish in shallow pools. Of course we could have gone to a roosting site during the day, but I much prefer seeing owls at night, when they are in their element.

We drive in an open safari 4×4 along the river. At every clearing we scan the treetops with a spotlight. The fishing owl's eyes should reflect bright red in the glow of the light. We have a false alarm twice; once for a nightjar and once for a Verreaux's Eagle-Owl, another imposing owl, with light gray plumage and unique pink eyelids that become visible when he blinks.

At the third attempt, we are lucky: Right in front of us, we see two red lights reflecting on top of a lonely little tree at the edge of a shallow pool. Our guide skillfully maneuvers the 4×4 closer. Then we turn the light back on. The bright red lights have turned into the dark eyes of a huge auburn owl. We have the chance to watch the Pel's Fishing Owl for a long time before he disappears from view with a few silent wing strokes.

Later that same night, I am awakened by a crackling sound outside my tent. I unzip the tent canvas and look straight into the eyes of an elephant, who is feasting on the young leaves of an acacia tree, less than half a yard away. It amazes me that an animal weighing more than 8,500 pounds is able to make so little noise. The elephant doesn't care about me, and I watch it quietly for several minutes. I can hear it breathe through the thin canvas and smell its typical zoo scent. It is so close that I can feel the warmth of its skin. What an exceptional experience! A wild elephant, so close. With only a wafer-thin canvas tarpaulin to protect me.

Johannesburg, South Africa

Camilla is already waiting for me when I walk through customs. She jumps into my arms and I hold her tight. It's been a bit of a struggle for the past few weeks, and the prospect of seeing her again kept me going during those moments.

"I'd love to hold you longer, but you really should shave off that filthy beard and take a shower first because you smell to high heavens," she says soberly.

Well, if she says so, because I have given up caring about it at all. Sometimes, I walk around in the same underpants for over a week, and my T-shirts are so worn out that you can see through them. No wonder the lady next to me on the plane held up her nose and gave me a disapproving look—she was, after all, seated next to a disheveled vagabond for an hour and a half. Fortunately, Camilla has brought me clean clothes and a razor so I can turn myself into a normal, neat Dutch chap again.

Ethan and his girlfriend, Billi, also came to the airport. The four of us will spend the next three weeks crisscrossing almost the whole of South Africa.

"I've gone through Noah's list again, and I think you can spot quite a few species that he missed!" Ethan had guided Noah around

Cape Town for a few days last year, so he knows better than anyone else where I can make a difference.

A little later, we find ourselves in a rental car on our way to the Suikerbosrand Nature Reserve. The sun is already low in the sky, and I have a hard time believing that we have enough time to find our target species, the Orange River Francolin.

Ethan reassures me: "When the sun has set, you still can easily continue to look for birds for another half hour before you really can't see anything anymore." He hits the gas pedal to back up his words.

Just before sunset we turn off the highway and drive on a small road into the reserve. Our rental car has a very low suspension and is clearly not made for this terrain, as its underbody scrapes along the raised median strip, making a terrible noise, as if ten people were scratching a blackboard at the same time.

"Don't worry," Ethan says. "It will not damage the car. Moreover, you have comprehensive insurance, so we could drive it off a cliff if we wanted to."

The landscape consists of rolling hills with endless grassland. Somewhere in this sea of grass, we have to find the francolins. Fortunately, Ethan met up with some local birders and asked if they could start looking.

We get out of the car at the highest point of the hill, where Ethan's friends are already waiting for us. And they have good news: "We just heard some francolins calling in that field by the road."

The sun has almost disappeared behind the horizon when we walk into the field. We stop every 10 yards and scan the knee-high grass with our binoculars. Suddenly I see them walking: three perfectly camouflaged, partridge-like birds. Only their heads are visible, the rest of their bodies hidden in the tall grass. We found them, and I give a thumbs up.

Ethan shakes his head: "We're not going to celebrate yet. We still have to find the African Grass Owl."

We walk to a swampy area at the bottom of the hill. It's really dark now. Ethan gestures for us to be quiet and plays the owl's call through his speaker. We continue to wait anxiously, but it remains silent. Second attempt. Nothing again.

"Okay, one last try then."

Suddenly the owl flies above our heads. One of the birders has his flashlight ready, and we can just see his heart-shaped, white face before he disappears into the darkness again.

Ethan looks at me triumphantly. "Now you can cheer."

Wakkerstroom, South Africa

It feels a bit like home away from home when we successively pass the cities Ermelo, Utrecht, and Amersfoort. Those names date from the seventeenth century, when South Africa was still part of the Dutch Cape Colony. We are on our way to the town of Wakkerstroom, in the middle of the Grassland Important Bird and Biodiversity Area. This area, which has been designated as one of South Africa's most important ecosystems, contains the greatest variety of endangered bird species in the whole country. You wouldn't guess that at first glance, because the landscape around Wakkerstroom consists exclusively of grassland.

Due to the centuries of cultivation of the landscape, we have a completely distorted picture of what a natural grassland should look like. Many people think the term *grassland* describes a green, trimmed meadow of ryegrass with a few stray dandelions and some grazing cows. That may look lovely, but such a place is in fact a completely human-controlled, sterile landscape, where it is virtually impossible for any form of life to exist without intervention. A healthy, dynamic grassland is rich in plant species and bursting with insect life. These areas are unfortunately becoming increasingly rare, due to the use of pesticides and the intensification of agriculture—not only in the Netherlands and in South Africa but all over the world. There are more than 8 billion people on our planet. All those mouths need to be fed, which makes humanity largely dependent on large quantities of land for agriculture, and so natural grasslands have become extremely rare.

Since 2001, BirdLife South Africa has been teaching young men and women from the surrounding townships to become bird guides. By involving local communities in bird tourism, BirdLife hopes to create support for the protection of this vulnerable grassland and the endangered bird species that live here. Ethan has arranged for us to go out with Lucky all day tomorrow. He has been involved in this project from the beginning and is the best guide of all. And someone with such a name cannot help but bring good luck.

It is early June, and, therefore, the middle of winter in South Africa. And the season is noticeable: The grass is barren and yellow, and at first sight there is not a bird to be seen in any fields or roads. However, we are counting our blessings, as it is a windless, sunny morning; it can be very cold here at this time of year, and sometimes there's even snow.

"Do you feel lucky today?" asks Lucky.

It's fortunate we have him with us, because all the grassland looks the same to us and we don't know exactly where to go. For us, finding a bird here is like searching for a needle in a haystack. Our target, the Botha's Lark, is a shy species that is difficult to find, even in spring when it sings. The world population of this small brown lark is estimated to be less than 2,500 birds.

We stop at a vast plain. I don't see any difference compared to the grassland we've been driving by for the past fifteen minutes. You wouldn't know it at first glance, but this area is intensively managed by BirdLife South Africa. It is seasonally grazed by sheep, and controlled fires are set once every two years. This way, they try to create the perfect conditions for the Botha's Lark.

We form a line, side by side, and walk into the grassland. According to Lucky, it is important to maintain this formation, with a gap of 10 yards in between each of us. "Otherwise the larks will scurry among us unnoticed."

Camilla clearly has trouble with this way of birding and is constantly distracted along the way by grasshoppers, plants, meerkats, and other beautiful things that pop up at her feet. You can't really blame her, as she's a biologist. Whenever she falls behind, we hear a frustrated scream from Lucky, who tries to keep her in line. More than an hour goes by: walking a hundred yards, a tirade from Lucky, and Camilla sprinting to join us again.

Then the redeeming words sound: "I have found them!"

A bit farther down, Ethan is beckoning me impatiently with one hand. With his other hand he holds his binoculars in front of his eyes. I run over to him and soon thereafter, I, too, see the perfectly camou-flaged pair of Botha's Larks, less than 5 yards away, scurrying like mice through the dry grass. Without Lucky's tactics, we would indeed have missed them. I give a thumbs up to Lucky; he has lived up to his name.

Durban, South Africa

This morning, we were unable to see even a single new species. I hate this, but Ethan hates this even more.

"It's not going to happen that you will have a zero day in South Africa," he says when we're at the airport, shaking his head. "Not on my watch."

Yesterday morning, we were still in the Drakensberg, at more than 3,000 meters altitude. From the land border with Lesotho—which is completely enclosed by South Africa—we could look out for miles over the lowlands. In this breathtakingly beautiful setting we saw the Drakensberg Rockjumper, a unique black-and-white passerine with an orange belly and rump and bright red eyes. This bird, as its name suggests, searches for food between rocks.

We spend long days in the field, resulting in only a few new spe-cies. I watch my lead over Noah shrink by the day and feel that haunting feeling of Madagascar slumbering beneath the surface again. But luckily, this time, Camila is with me, and I can put every-thing into perspective: I'm in the midst of the adventure of my life, with my great love by my side, so there is no reason whatsoever to throw in the towel.

When we arrive in Cape Town, it is ten o'clock in the evening and pitch dark. I can't see any new owl or nightjar species here anymore, so it looks like today will really go down in the books as my second day with zero new species. But not if it depends on Ethan. During the flight to Cape Town, he came up with a plan to still see a new species today. His idea is to drive to the coast as fast as possible and then shine a flashlight on the beach.

"We will have a one hundred percent chance to see a Hartlaub's Gull, because they are extremely common around the Cape."

No sooner said than done. After picking up our rental car, we drive to the coast. I point my flashlight at the beach and, to Ethan's dismay, we don't see any Hartlaub's Gulls. We quickly drive on to another beach and find no gulls there, either. Meanwhile, time is running out. We still have fifteen minutes until midnight, so we can make one last attempt.

There we go; it will be all or nothing. I grab my flashlight and shine it from left to right across the beach. Nothing but sand.

"Move back a little. I think I saw a bird," Camilla says.

Slowly I move the beam back to the left. She is right. There is a lonely bird on the beach. I grab my telescope. It's not a Hartlaub's Gull but something even better—an African Oystercatcher, a beautiful jet-black wader that is found only along the South African coast. Mission accomplished!

Cape Town, South Africa

I now belong to a higher cult of mortals, for I have seen the Albatross.

—ROBERT CUSHMAN MURPHY (1912)

Albatrosses are arguably the ultimate birds. They roam the oceans their whole lives, coming ashore only to breed. With their perfectly aerodynamic physique and enormous wingspan, they soar effortlessly through the troughs of waves, barely needing to flap their wings and reaching speeds of over 75 miles per hour.

They grow the oldest of all birds. A famous example is a Laysan Albatross named Wisdom, the oldest known wild bird. This year, again, this albatross granny managed to raise a young at the age of sixty-five (when the English version of this book will be published, she will have done it again, at the age of seventy-one)! She is recognizable by a color ring on her left leg, and she returns annually to the same seabird colony in the Midway Islands, where she herself hatched in 1951.

There are twenty-two species of albatross in four genera. The genus *Diomedea* grows the largest in size. One species within this genus, the Wandering Albatross, can reach a wingspan of more than 11.6 feet, larger than any other bird. To put this in perspective, a Bald Eagle has a wingspan of no more than 6.5 feet.

As many as seventeen of the twenty-two species are endangered, and some of them are critically endangered. The main causes are pollution and overfishing. Albatrosses are dying in fishing nets and hooks from long-line trawlers, and in some colonies, alarming amounts of plastic have been found in their stomachs. Since most species are sexually mature after six to ten years and raise only one young every two years, these factors have huge impacts on the populations of these beautiful birds.

There are only a few places in the world where you can see large numbers of albatrosses on a day trip at sea. Cape Town is one of those places, due to a rare convergence of circumstances. Off the South African coast lies the Agulhas Bank, a submarine plain sloping from 60 to 300 feet deep. This plain ends abruptly 30 miles off the coast, with a precipice of more than a half mile straight down. This underwater cliff is where the cold Atlantic and warm Indian Oceans meet, causing nutrient-rich water to well up from the depths. This causes an explosion of marine life, from microscopic plankton to 60-foot-long sperm whales.

Unfortunately, this marine life is threatened by overfishing. Devastating miles of trawl nets destroy the seabed and systematically drain the waters around Cape Town. This is not just a South African problem but a global one. In recent years, the breeding success of the endemic African Penguin, the Cape Gannet, and the Bank Cormorant has been alarmingly low; there is simply not enough food to raise their young. If we want to save these species from extinction, we need to switch to a sustainable form of fishing, which gives marine life a chance to recover. We all have to work together on this. If you buy fish in the supermarket, first check whether it has an eco-label certification for sustainably caught fish.

———◦—◦———

It's five o'clock in the morning and today is the big day: We are going to make a pelagic trip.

Luck is on our side, as the weather is good enough to set sail. It can sometimes become quite hairy off the Cape coast, and usually more than half of the trips are canceled.

Our trip is organized by Cape Town Pelagics. This nonprofit gives 100 percent of its proceeds to seabird research and conservation and is a major donor to the Albatross Task Force, a BirdLife conservation campaign designed to halt the catastrophic decline of seabirds. They have already had some great successes. For example, it is now standard in longline fishing to weigh down the fishing lines and attach brightly colored ribbons above the cables. The weights prevent the albatrosses and other seabirds from reaching the fishing hooks, and the ribbons act as a deterrent, preventing the birds from flying to their death against the cables. These two simple, cheap measures have reduced the bycatch of seabirds by 90 percent and of albatrosses by 99 percent.

———◦—◦———

We set sail very early in the morning, on a small motorboat from Simon's Town harbor through False Bay toward the Cape of Good Hope. The weather is nice and sunny, and the sea is calm, or so it seems.

"Just enjoy these calm waters for a while," says Clifford, our guide and boatswain. "As soon as we get out of the bay, the party will start."

By this he refers to the ocean swell, several yards tall.

Camilla looks at me with frightened eyes. She has anything but sea legs, and during previous boat trips on the North Sea on her father's fishing boat, she invariably ended up sick in the foredeck.

"Just concentrate on the horizon and everything will be fine," I assure her.

In front of the Cape of Good Hope, Clifford stops the boat for a while. The lighthouse on top of the cliffs is shrouded in mist, and waves crash against the rocks far below. It's a dramatic sight. I am reminded of the countless sailors who have come to their end here. What a horrible experience it must have been to round the Cape in the seventeenth century during a storm, your teeth falling out from scurvy, while huge waves crash against the hull of the wooden VOC (Dutch East India Company) ship.

We head south. Soon we find ourselves on the high seas, and the Cape has been reduced to a speck on the horizon. The same horizon regularly disappears from view for several seconds at a time due to the ocean swell. I see Camilla turn white.

Then Clifford shouts, "Albatross! Portside!"

A moment later, a Black-browed Albatross glides past the boat at a distance of merely 20 yards. We can follow the bird for several minutes, and all the while it doesn't have to flap its wings even once. Being this close, we can clearly see its light orange beak and dark eye stripe.

"What a great sight, don't you think?"

Camilla raises her thumb but immediately turns green and manages to find the railing just in time. Luckily for her, she's not the only one. Three other participants soon follow suit and I, too, have great difficulty suppressing my nausea. But for everyone on the boat, seeing albatrosses is a dream come true, so we bravely press on.

As we head farther out to sea, the number of seabird sightings increases. We see White-chinned and Cape Petrels, Antarctic Prions, and Wilson's Storm Petrels. The last are not much bigger than a starling and regularly paddle on the water, allowing us to occasionally see the bright yellow webs between their toes.

Then Clifford starts pointing and shouts excitedly, "There's a fishing boat out there, pulling in its nets!"

He turns the wheel. We bounce over the water at 25 miles an hour, and the few who were not yet seasick are now vomiting just like the rest. Ten minutes later we have reached the fishing boat. The pulleys at the back of the fishing boat are running at full speed to raise the net with heavy steel cables from the depths. Thousands of opportunistic seabirds swim and fly behind the boat, hoping to get their share. I have never seen anything like it. Four species of albatrosses fly crisscross over one another and sometimes skim just past our boat.

I notice that a few yards above the steel cables, two more lines are fluttering with long, bright orange ribbons. A Shy Albatross flies straight toward the cables, but at the sight of the orange ribbons, it swerves at the last moment and lands safely on the water a little farther away.

Just before we have to return to the port, Clifford suddenly starts shouting: "*Grey-headed Albatross!* Portside, right next to the boat!"

Two yards away, an albatross with a gray head and a black beak is bobbing on the water. We can't believe our eyes: This is South Africa's first sighting in five years. We go completely crazy, and even Camilla, who has donated her entire stomach contents to the sea, gets a bit of color on her face from pure excitement.

Accra, Ghana

Early this morning, I have to say goodbye to Ethan, Billi, and of course Camilla. This goodbye to Camilla is a lot less difficult than last time, because in less than two weeks I have another stopover in the Netherlands, and then the two of us will travel through Spain for three days.

My count currently stands at 3,401 species. In South Africa, my list grew slowly, but we actually didn't do so badly. In fact, during my nineteen days in South Africa, I observed more than forty species that Noah had missed during his visit, while in eighteen days he had only three species that I did not manage to see on my trip.

Up till this point, that one day in Tanzania is still the only day with zero new species, but that could be about to change.

It is almost dark and pouring with rain when I walk out of Accra airport. I am met by two enthusiastic Ghanaian men with binoculars around their necks and a sign that reads ASHANTI TOURS WELCOMES ARJAN DWARSHUIS. They introduce themselves as Paul and William. I explain to them that we have no time to lose, whereupon they look at me as if I've gone mad.

I tell them that a day with zero species could give people the impression that Ghana is not as rich in birds as everyone always says.

My observation hits them like a brick. They look at me in bewilderment. "This cannot happen! Ghana is great!"

A few minutes later, we are standing in the pouring rain at a field next to the airport. We have only fifteen minutes before it will really be too dark to see anything, so we have to be quick. A bird with a long tail flies in and lands a little farther on the fence. Through my binoculars, I can see that he has a striking black mask and a yellow beak: a Yellow-billed Shrike. A new species!

We breathe a sigh of relief and settle down on a terrace nearby. We are waiting for a flight from the Netherlands, which is delayed by two hours. So I order some food and a few beers to celebrate the sighting of the shrike with Paul and William.

Finally the flight has landed. The first to walk out is my father, followed by Michiel and John.

During the two-hour drive to Kakum National Park, my father proudly shows me his handmade photo guide to Ghana. He found photos of the more than seven hundred species that occur in the country and bundled the end result in a folder. What a painstaking job. He found the photos via Google, so there are some mistakes. Moreover, his efforts were not strictly necessary, as we carry the book *Birds of Ghana* with us. But still, it touches me that my father supports me all the way in chasing my dreams.

⁓

Kakum National Park protects 145 square miles of West African lowland rainforest, which is home to dozens of endangered animal and plant species. It is known for the Kakum Canopy Walkway, an ingenious construction of rope bridges connecting seven wooden tree platforms. This dizzying hiking trail, dozens of yards high, offers visitors who don't suffer from vertigo a unique view of the crown of the rainforest. Birders also have the chance to see, at eye level, a number of species that live exclusively between the treetops.

That night we sleep in a tree house in the middle of the park. Michiel arranged that for us.

"That adventurous note is good for the viewing rates, which is why *Expedition Robinson* scores so high. People want to see you suffer," Michiel adds.

We can count ourselves lucky. In Celebes, he had arranged a plastic sheet in a mosquito-infested jungle as a place to spend the night. Compared to that, this tree house is a five-star hotel.

Paul and William ordered some pizzas from the little restaurant at the entrance to the park, then transported them vertically in a plastic bag to the tree house. Both our dinner that evening and our breakfast the next morning will consist of a greasy heap of dough with several different toppings.

After a breakfast of cold pizza, we walk in the dark to the Canopy Walkway. The tree house has one big advantage: It is located in the middle of the park, so we can get onto the walkway at first light, before the park officially opens up to visitors.

The horizon slowly turns red as we shuffle over a wobbly rope bridge toward the first platform. The structure is built against a steep slope, and the observation platforms are attached to enormous, 50-yard-tall forest giants that tower above the rest of the forest. We soon find ourselves above the canopy.

We continue to the third and highest platform, from where we have a phenomenal view over the rainforest. Fragments of morning mist hang between the treetops. As soon as the morning light slowly breaks through, the forest comes to life.

The first birds we see are five huge, black-and-white Brown-cheeked Hornbills. They fly by at a distance of about 50 yards and land on the top of a tree, where they start pecking ripe wine-red fruits with their light yellow beaks. Shortly thereafter, William notices the distinctive hyena-like call of a White-crested Hornbill. He plays the sound through his speaker, and we don't have to wait long before a pair of this impressive hornbill species lands on a bare branch below us. Through the telescope we can study their long, black-and-white tails and bushy white crests in great detail.

The highlight of the morning is a huge mixed flock passing through the treetops almost within touching distance. I see a number of species that until that moment I only had the opportunity to see from below, looking up with a stiff neck from a considerable distance.

Despite the fact that our visit to Ghana falls in the middle of the rainy season, all day long we have sunny, windless weather. This allows Michiel and John to film in peace, without having to worry about soaking wet camera equipment. This is a relief, especially for John, because the weather conditions in Celebes had been far from ideal.

Bonkro, Ghana

My great hero is Sir David Attenborough. He is the epitome of natural history and has inspired millions of people around the world with his enchanting voice and charisma. He had a long career with the BBC, where he presented impressive wildlife documentary series, such as *The Living Planet* (1984) and *The Life of Birds* (1998).

His on-screen career began in West Africa in the early 1950s. At that time, he was working for a television show, *Zoo Quest*. In this BBC documentary series, Attenborough followed curator Jack Lester on his expeditions to capture new species for the London Zoo—in the mid-twentieth century, capturing wild animals and shipping them to zoos was still a normal thing to do. The plan was that Lester would present the series, while Attenborough would be the producer, but when Lester fell ill (a few years later he would die from an unknown tropical disease he contracted during one of his expeditions), Attenborough took over the role of presenter.

During the recordings, they went to search for the White-necked Rockfowl, also known as *Picathartes*. The black-and-white images showing Attenborough visiting the nesting site of this spectacular bird, deep in the rainforests of Sierra Leone, went around the world, kicking off his impressive career.

Attenborough, who nowadays is an avid animal lover and conservationist, still regrets being involved in the capture of wild animals. "But those were different times," as he puts it. And he is right about that. Fortunately, he more than made up for this "mistake" and even in his nineties, still presents nature documentaries and is one of the world's leading figureheads of nature conservation.

The White-necked Rockfowl leads a secluded life in the tropical rainforests of West Africa, from Guinea to Ghana. Its only close relative is the Grey-necked Rockfowl, which occurs farther east in West Africa, from Nigeria to Gabon. Together they form the Picathartidae family, which is believed to have emerged about 44 million years ago. Both species are considered to be vulnerable because the rainforest in West Africa is being cleared at a rapid pace.

The ecology of the *Picathartes* is unique. They live in small flocks in the rainforest, where they hop on the ground in search of food. They have short wings and are, therefore, poor fliers. A pair of *Picathartes* stays together for a lifetime. They make bowl-shaped nests of mud and plant fibers under a gigantic overhanging rock near a stream in the forest. These nesting locations are scarce and, therefore, often used by several pairs.

At least as unique as its breeding ecology is its appearance. A White-necked Rockfowl is the size of a small crow, with a snow-white neck and belly, slate-gray wings and mantle, and a long, gray tail. It has a bizarre, bright yellow, bald head, with two conspicuous black spots behind both eyes, which makes him look like he's wearing some kind of latex mask.

For a long time, it was thought that the White-necked Rockfowl had become extinct in Ghana, as there had been no reliable observations since the 1960s. But in 2003, the species was rediscovered by a team of American ornithologists in a community forest reserve near the remote village of Bonkro. There appeared to be several nesting locations in the reserve. After the researchers had studied the birds for several years and determined that there would be no disturbance, one of these colonies was opened up to the public. The local community of Bonkro was closely involved in this operation.

Just before entering the village, we stop at a construction site along the road. A sign reads ASHANTI AFRICAN TOURS PICATHARTES CONSERVATION AND COMMUNITY DEVELOPMENT.

A group of men is pouring a cement floor. They interrupt their work to greet us. One of them says that they are building a community center and a school. Anyone who visits the reserve to see the

Picathartes has to buy an entry ticket. The revenue from these entry tickets help to finance this construction project, which is sponsored by Ashanti African Tours. It also encourages the local community to protect the forest and the animals that live there.

We receive a warm welcome in Bonkro. The women wear beautiful, colorful skirts, and the men have put on their neatest suits. There is music, and a group of boys and girls from the village put on a dance performance. Michiel slyly pushes me forward so that I cannot but dance along; there are not many things I dislike more. John is instructed by Michiel to capture my awkward dance steps and my red, blushing face.

When the festivities are over and we have paid our entry tickets, we leave the village with two young guides and head for the community reserve. We walk in a row on a narrow path through banana and cocoa plantations. A little later we arrive at the edge of the forest. The trail continues through pristine rainforest. It's noon, but it feels wonderfully cool in the woods, as the scorching sun is blocked by the dense canopy. In the background, we hear the constant, monotonous hum of cicadas and the hollow song of a Black-throated Coucal.

We come to a halt at the bottom of a stony slope, in the middle of the forest.

Paul turns and addresses the group: "From now on, everyone has to be quiet."

We climb up the slope as quietly as possible. I'm tense; the *Picathartes* nest can't be far away now. We stop at a small clearing at the foot of a gigantic overhanging rock and take a seat on a wooden bench. Paul grabs my arm and points to the rock wall. About 20 yards away, I see a bowl-shaped structure of dried mud stuck to the overhanging wall. It is the size of a shoebox and features a crescent-shaped hole on one side: I am looking at the nest of a *Picathartes*.

And now the waiting begins. An hour goes by, and we haven't seen or heard anything yet. I wonder if Attenborough ever found himself in a similar situation and if he would have felt the same way sixty years ago in Sierra Leone.

Suddenly, the bushes at the edge of the clearing start to move, and before we know it, a magnificent White-necked Rockfowl jumps out. My heart skips a beat. For a while, the bird stands motionless on a fallen tree trunk, and I cannot but watch it breathlessly. Its plumage is so perfectly white and gray that it's hard to imagine that this is a bird that lives on the muddy soil of the rainforest. There's something cartoonish about the bald, yellow-and-black head. I've never seen a bird that even remotely resembles this one.

A second bird jumps out. In the hour that follows, the *Picathartes* couple take turns sitting on the nest, while we can observe them without disturbing them. It also gives me the opportunity to follow in the footsteps of my great hero: Squatted in front of the camera, I talk about the *Picathartes* while the actual birds are part of the scenery in the background. And for a moment I imagine myself as a young David Attenborough.

Sierra de Gredos, Spain

"I see a White-throated Dipper!"

I run to Camilla, and a few seconds later we are watching the bird together. He sits on a stone in a fast-flowing stream, bobbing up and down relentlessly. It's because of this compulsive behavior that this black-and-white bird is called "dipper." Every now and then, he dives from the stone into the water to resurface a yard farther down. Dippers are the only songbirds that are able to swim underwater. They shield their eyes with a sort of transparent membrane, the bird equivalent of goggles.

———

Today is July 3 and we are in the mountains of central Spain. Four days ago, I arrived at Schiphol, after traveling nonstop through Africa for almost three months. Only my parents, Camilla, and a few close friends knew about my lightning visit to the Netherlands, because there was a lot of work to be done and the next afternoon Camilla and I would fly to Seville.

I repacked my bag, ate a Hollandse Nieuwe, which is raw and lightly salted herring, at my favorite herring stall in Scheveningen, slept in my own bed, and sighted almost all the northwestern European bird species that were still missing from my list.

After our dipper sighting, we drive to the Madrid airport. While I still can enjoy Camilla's company for a little while, I'm painfully aware that we will be separated again for two months, until we reunite in Peru. In the meantime, I will visit Puerto Rico, the Dominican Republic, Suriname, Brazil, Argentina, and Chile.

I'm in the mood for South America. The Old World (Europe, Asia, and Africa) and the New World (North, Central, and South Americas) have hardly any bird similarities, so almost every species I will see in the period ahead will be new. And that's nice, because in the past month, the results have sometimes been meager. Especially in Tanzania and in the northeast of South Africa, my count crept up excruciatingly slowly, which did not exactly help to lift my mood. I noticed that sometimes I was more preoccupied with my species list than actually enjoying the birds themselves and the beautiful nature surrounding them. Fortunately, I am now well on schedule. The prospect of almost four months of intensive birding in South America raises my spirits, and I feel completely ready for it again.

When I arrive at the check-in desk, everything seems to be okay. My flight to Puerto Rico will leave on time, and I still have more than an hour to stroll to the gate.

"You don't have a valid visa, sir."

Did I hear this right? A visa? Why on earth do I need a visa for Puerto Rico, a Spanish-speaking island in the Caribbean?

"Don't you know that Puerto Rico is part of the US?"

This hits me like a ton of bricks.

"No, I didn't know."

Immediately panic sets in. My flight leaves in less than an hour, and in that time I have to get a visa for the US, which normally takes up to several days. I feel the blood drain from my face. In my mind, I see my itinerary go down like a house of cards. If I don't arrive in Puerto Rico today, I'll miss my birding day there and possibly even my connection via the Dominican Republic to Suriname, as this combination of flights is available only twice a week at most. This blunder could cost me a hundred species.

Camilla's flight to the Netherlands also leaves in an hour, and that means I'm on my own. I decide to call my parents; maybe they can arrange something at home on the computer. When I get them on the phone, I have exactly forty minutes until my gate closes. Fortunately, they remain calm.

"Give us your passport details and we'll see if we can get something done."

Meanwhile, Camilla has to run to catch her flight.

"I'm so sorry to have to leave you here, honey. But all will be fine and we'll see each other in two months." She kisses me and leaves me behind at the check-in desk.

Fifteen minutes later my phone rings.

"I think it worked," my mother says.

How? There is no time for an explanation, and I sprint to the desk.

"Your visa is okay, but unfortunately the gate is closed."

"But I still have fifteen minutes!"

"I'm sorry, sir."

I sink down to the floor with my back to the desk. What now?

I decide to call the twenty-four-hour help desk of my sponsor ATPI. Less than ten minutes later, they ring me back. They've found a new flight, from a different terminal, departing in exactly one hour.

I decide to take the gamble and sprint toward the shuttle bus, which I just manage to catch. Half an hour later, I arrive at the check-in desk, pouring with sweat. I'm just in time: The next flight to Puerto Rico wouldn't leave until eighteen hours later. I give a sigh of relief. The meal on the plane has never tasted so good.

Due to the time difference of five hours with Madrid, I arrive in Puerto Rico at eight PM that same day, in the dark. My guide is already waiting for me. "Put your bag in the car quickly," he tells me, "because we have only tonight to find the Puerto Rican Owl."

An hour later, we stop at a park. When I get out of the car, I can already hear his call. This species can't be found anywhere else in the world. We don't have to look long with our flashlights until we find him: a brown owl with bright yellow eyes, looking at us from a bare branch.

What a crazy day this has been.

Amsterdam, Netherlands (2015)

It was early October, just under three months before the start of my trip, when the telephone rang: "You are talking to Anne from *RTL Late Night*. Would you like to come on our show to talk about your world record attempt?"

"Of course, I would like that."

I could really use the publicity.

"There is one condition: The host, Humberto, wants to go birding with you one of the mornings before the broadcast."

A few days later at seven in the morning, I was waiting for Humberto Tan in the dunes of Meijendel. I was slightly nervous, as Humberto had been voted the best-dressed man in the Netherlands for several years in a row, and I expected that he would turn up in a three-piece suit and calfskin shoes. Nothing turned out to be less true; he got out of his car in wellies and a rain suit, with a camera crew in tow.

"Wow, you have a posh scope!" He looked at my Swarovski telescope with admiration.

Bird migration was at its peak, and the sea buckthorns and elderberries in the dunes were full of thrushes and Eurasian Blackcaps. We stopped, and I pointed out the many bird sounds around us.

"Firecrest, Short-toed Treecreeper, Coal Tit, Common Chiffchaff, Song Thrush, Common Chaffinch." I quickly fired names of species at Humberto, explaining which bird belonged to which sound.

"How do you know all this?"

I explained that I am always subconsciously watching the sky and listening for any bird sounds, even when I am in a conversation.

He was highly amused hearing this and imagined what it would be like if he also suffered from that disorder: "I can already see myself interrupting an important live interview because out of the corner of my eye I see a buzzard or I hear the song of a Great Tit."

I suggested we take a look at the Meijendel bird ringing station.

"What is that, a bird ringing station?"

I told him that thousands of birds were caught here every year, given a ring, and then released again. If such a bird is recaptured or found dead elsewhere, we can learn a lot about its migration route and its distribution.

"Good morning!" Vincent was standing next to a 30-foot-long mist net that was set up along the bushes, smoking a cigarette. From where we were standing, I could already see that three Goldcrests, a Short-toed Treecreeper, and a Chiffchaff had been caught in the net.

"This Goldcrest weighs about five grams," Vincent said after carefully removing the bird from the net. "That's about as much as a teaspoon of sugar."

After explaining to Humberto how to hold the bird, Vincent gently placed the Goldcrest in his hand. He then blew aside some feathers on the bird's belly, revealing two yellow spots on its breastbone.

"Look, he's got a lot of fat, which means this Goldcrest is in top condition."

Humberto stood open-mouthed, staring at the tiny bird in his hand, after which he canceled all his appointments for the morning. We stayed at the ringing station for another two hours and saw Eurasian Blackcaps, Redwings, Song Thrushes, and even a Firecrest.

Humberto became a birder that same morning.

When I joined his talk show a few days later to talk about my Big Year, his enthusiasm had not waned. This allowed me to tell my story in a relaxed way, live in front of nearly a million viewers.

A few weeks later, we met up again to go birding. While walking through the Amsterdam Dunes on a drizzly November day, we talked about the adventure that awaited me.

"I'm a little jealous of the crazy trip you're about to undertake."

"Then why don't you join me for a while?"

I realized this was next to impossible for Humberto. He had to present his talk show until June and then report on the Olympic Games in Rio de Janeiro less than a month later.

"Are you also going to visit Suriname by any chance?"

I told him I would be there the second week of July.

Two weeks later he sent me a WhatsApp with a screenshot of his ticket from Amsterdam to Paramaribo.

Paramaribo, Suriname (2016)

"Hey, dude, are you ready for it?"

In the doorway of my hotel I see Sean, the best birding guide in Suriname. I met him seven years ago, in the remote rainforest of Kabalebo, at the border with Guyana. We spent a few days birding and I promised him that, one day, I would return to his beautiful country.

After I quickly splash some water on my face, I walk to the breakfast room, where I am greeted by my travel companions: Michiel, Humberto, and the Surinamese birder Fred Pansa.

My count now stands at 3,646 species, with just under half a year left. For the next three and a half months I will traverse South America, the most bird-rich continent of all. When I leave this continent at the end of October, I hope to have passed the magical number of 6,000 species.

Suriname has less than 600,000 inhabitants, while it is almost four times the size of the Netherlands. This makes it one of the most sparsely populated countries in the world. More than 90 percent of the country consists of tropical rainforest. Suriname is a paradise for birders; even the suburbs of Paramaribo are bursting with birds.

While the sun slowly rises, we drive in a van over the 170-foot-tall Jules Wijdenbosch Bridge to the west. We see the Suriname River far below us.

The Peperpot Nature Park is situated along the riverbank. It used to be a plantation with slaves until the mid-nineteenth century.

When slavery was abolished, the plantation passed to contract workers, and Peperpot remained in use until 1996. Then it became overgrown with greenery. Over the years, wildlife has slowly returned. Camera traps have confirmed the presence of jaguars, cougars, tapirs, and giant anteaters.

As soon as we get out of the car, we hear a symphony of birdsongs: More than 400 different species have been identified in this 1,700-acre park.

We walk into the forest and, immediately, bird activity erupts in full force. After we have walked the first hundred yards, I have already seen more than twenty new species. Even for me, it's hard to keep up; but for Humberto, who has called himself a birder for only nine months, it is as if he is thrown into the deep end without a swimming certificate.

I point out a bird in the treetops.

"There, up in that tangle of vines, a White-flanked Antwren!"

"A what?"

I see him looking through his binoculars. Then he looks at me with a helpless look: "A White-hand Chicken? Look, I don't see it, nor do I understand it."

Fortunately, Humberto turns out to be a quick learner, and by the end of the morning, when we leave Peperpot, he has already figured out the more ineffable art of birding.

We continue our journey to the Brownsberg Nature Park, an inland nature reserve a few hours south of Paramaribo. The road is teeming with birds, and every stop brings a few new species. As we arrive at Brownsberg at dusk, a Short-tailed Nighthawk hunts overhead for insects, my 113th new species today, which makes this the most productive day since my first day in the Dutch Rhine Delta.

Our accommodation is located in a clearing in the middle of the rainforest, on top of a hill at an altitude of 400 meters. It offers beautiful views over the Brokopondo Reservoir, which lies far below us.

In the morning, the clearing is visited by a flock of Grey-winged Trumpeters. These large, fowl-like birds, with glossy black plumage and striking light gray wings and rump, are distant relatives of cranes.

They are photogenic animals, and Humberto spends hours trying to take the perfect picture.

There is a network of forest tracks, which makes for delightful birding; we never have to look long for a mixed flock, which sometimes consists of dozens of different species. The highlights follow each other in rapid succession, and in the evening, when we have a beer in the dining room, I feel exhausted but satisfied. It is hard to let everything sink in. Everything seems to be going well, maybe a bit too well.

On the morning of our second day in the Brownsberg, Michiel wakes up drenched in sweat. He looks pale and has deep bags under his eyes. An hour later, he is shivering with cold, while it is 86°F. He's clearly not well, but he thinks it's nothing serious and blames it on an allergic reaction to sand flea bites, something we both suffer from—our legs are covered in dozens of purulent pus bubbles that itch terribly. Michiel is anything but a poser, and he insists we continue birding.

He starts to look worse and worse, and I get seriously worried. Suddenly, the penny drops. Less than two weeks ago we were in Ghana. During the rainy season, malaria is a major problem there. All his symptoms point in that direction: high fever, chills, headache, and profuse sweating.

We decide to drive to a nearby field hospital, under fierce protest from Michiel because he prefers to continue birding. When we get there an hour later, he is shivering in the back of the car in a fetal position.

"You have *malaria tropica*," concludes the doctor, after doing a quick test. This news hits us like a ton of bricks: This is the most deadly form of malaria.

"Can you quickly get me a cold Coke from somewhere? I feel I'm going to faint." Michiel looks at me with red-rimmed eyes and is as white as a tablecloth.

We decide unanimously that he has to go to a hospital in Paramaribo immediately.

"Order a taxi for me, because you guys have to get on."

A little later, Michiel leaves for Paramaribo in the back seat of a taxi. I immediately feel guilty. His white face in the back of that taxi will be engraved in my mind forever. I should have gone with him, Big Year or not. He would have done the same for me.

In retrospect, he will say he didn't blame me, yet this decision still doesn't feel right: I gave priority to my record over the well-being of a good friend. If things had gone completely wrong and he had died, I would never have been able to forgive myself. This whole adventure and the world record wouldn't have meant anything anymore . . .

Zintete Lodge, Little Saramacca River, Suriname

Fred Pansa comes from a family of twenty-six siblings. His father is the tribal chief of Apiapaati, a Maroon settlement on an island in the Suriname River, deep in the interior of Suriname.

The ancestors of the Maroons were West Africans who, from the sixteenth century, were forcibly brought to South America by slave traders. They managed to escape from slavery and founded small settlements deep in the jungle.

The Surinamese Maroons waged, successfully, a guerrilla war with the white settlers. As a result, various tribes were given legal freedom and an autonomous status in the mid-eighteenth century. Over the centuries, they developed a West African culture outside of Africa.

Fred grew up as a child of the forest, and the Amazon rainforest was his backyard. As a child he always had a great interest in birds, but he had never heard of the term *birdwatcher*. He was very ambitious and, at a young age, started organizing inland tours to introduce tourists to the Maroon culture.

A few years ago, he had led a similar tour and was on the plane back to Paramaribo. As he stared out of the window at the endless green blanket of treetops below, his attention was caught by a barren granite mountain, towering over the surrounding landscape like a huge bulge. A little farther down, he saw the beginning of a dirt road. He pressed his nose against the window and saw that this small road ended at an intersection with a tarmac road.

Fred became obsessed with the mysterious mountain and decided to investigate it. There are not many roads in Suriname that go inland, and when he plotted his flight path on a topographic map, he concluded that the asphalt road could only be the road from Paramaribo to the south of Suriname. He remembered that the small dirt road bridged a river. Not far from the point where his flight path and the asphalt road crossed on the map, there was indeed a small river: the Little Saramacca River.

And so, armed with a machete, he began his quest. He soon found the crossroads. The dirt road turned out to belong to an old logging concession. There were no trails at the end of the road, so he had to cut his way to the top of the mountain. This turned out to be quite a job in the murderous tropical heat. After three failed expeditions, he finally succeeded at his fourth attempt. At the end of December 2013, he stood on top of a granite rock, from where he had an overwhelmingly beautiful view over the Surinamese rainforest. The mountain didn't have a name yet, so he named it Fredberg.

He hasn't stopped ever since. He carved a clearing along a tributary of the Saramacca and started building a lodge, surrounded by a network of forest trails. He has also built an overnight accommodation on top of the mountain. In only a few years, he became an accomplished bird expert. The Fredberg bird list now stands at more than 400 species, and new birds are added to this list almost every week.

Topping that list is the Harpy Eagle, South America's largest and strongest eagle and the holy grail for any birder visiting the Amazon rainforest. Fred frequently sees this mighty bird of prey in the vicinity of Fredberg, and I have the silent hope that we will see one during our three-day stay.

We are some of the first birders to visit Fred at the Zintete Lodge. He proudly shows us his plans: "I want to build a veranda here and toilet cubicles will be over there. The dorm will get a second floor, and I also want to build a bar."

The dorm is still under construction. If we want to take a bath, we have to dive into the river with a bar of soap. But seeing what Fred has accomplished in such a short period of time, I have full confidence in the final result (since our visit, the Zintete Lodge has grown into one of the best-known birding lodges of the Guiana Shield).

We sleep in hammocks stretched out between the walls of the dormitory. The lodge is not yet equipped with a generator, so we have dinner by candlelight. Sean plays Surinamese songs on his guitar, and every now and then we hear the call of a Spectacled Owl coming from the forest at the edge of the clearing. There is no mobile network, so we are not distracted by our cell phones. Moments like that are rare these days, especially for Humberto, who is used to the spotlight of show business. We can only relax, listen, and enjoy the perfect tranquility around us. Yet, I feel awkward. My thoughts are with Michiel, all on his own in a hospital in Paramaribo, fighting against malaria.

Sean is behind the wheel and Fred is on the roof of the van, so he can keep an eye on the treetops along the road. It is late morning, and we drive around at walking pace in the vicinity of the lodge. Suddenly Fred hits the roof with his fist. That can mean only one thing: He saw something!

Sean hits the brakes, and we jump out of the van. Almost immediately we see a gigantic bird of prey, sitting in a dead tree less than 50 yards away. The bird has a light gray head with a fluttering crest and a broad black breast band. Its belly is white and it has huge pale yellow talons with dangerous-looking black nails that would allow it to pierce the skull of a sloth or howler monkey with playful ease. We don't have to doubt for a second about the identity of the bird: It is a Harpy Eagle. Trembling with excitement, I grab my telescope. When I zoom in on his head, I have goosebumps on my arms. His inky black eyes and huge, hooked beak give him an almost demonic appearance. What an imposing animal! Slowly he turns his head from one side to the other, taking a close look at his surroundings.

He doesn't seem to care about our presence; he doesn't even look at us. Then his eyes fix on an invisible point on the horizon. He takes off and, with a few mighty wing strokes, flies straight over our heads into the jungle, perhaps for a meeting with an unfortunate howler monkey or sloth.

On our last day at the Zintete Lodge, we climb Fredberg. It is a tough but extremely beautiful 3-mile walk, straight through the rainforest. It's hard to imagine that, two and a half years ago, Fred made this trail all on his own, with only a machete in hand.

The forest is pristine, and sometimes we see tracks of tapirs and giant anteaters on our path. Twice we find jaguar paw prints, one of which, according to Fred, is no more than half an hour old. Humberto and I look around uneasily and start to walk a bit quicker. That big cat might just be hiding in the impenetrable vegetation, ready to prey on someone that lags behind.

Fred had not exaggerated: The view from the top of Fredberg is fabulously beautiful. In all directions I can see only greenery, as far as the eye can see. Every now and then groups of colorful macaws and parrots fly low over the treetops, and in the distance we hear the unmistakable "*Huui É T-wiew!*" of a Screaming Piha. Reminiscent of a construction worker whistling at a beautiful lady, this sound can be heard in almost every Vietnam War film—a classic production flaw, as this songbird can be found only in the rainforests of South America.

To our surprise, Sean has a signal on his phone, and we decide to call the hospital in Paramaribo to check in on Michiel. We get him on the phone and he tells us that the medication is working and he is on the mend. What a relief!

With 328 new species in just seven days, Suriname is the most productive country of my Big Year so far. My count stands at 3,974 species, and I am leaving for Brazil in good spirits.

Boa Nova, Brazil

"I hear an owl calling," Ies says while he's unloading our luggage. Ies is a fanatic Dutch birder and a world traveler, just like me. We always had the plan to go on a big birding trip together, and when he heard that I was looking for travel companions for my Big Year, he didn't have to think about it twice. "I'm going with you to Brazil and Argentina," he told me with a broad grin on his face during the Dutch Birding Association Day.

I'm standing at the reception of our hotel with a sleepy head. It is the middle of the night, and we have just driven nearly 300 miles. After a long flight with three stopovers, I had arrived in Salvador, on the east coast of Brazil, that very afternoon.

"That owl call can't be anything exciting," I shout back.

All I can think about right now is sleep. After some insistence from Ies, I reluctantly trudge outside. Now I hear it too, a soft "*Hoé*," repeated every few seconds. I get out my flashlight and soon we have found the bird, which is perched on a palm tree in the middle of the town square. It is a remarkably large, dark owl, and I can't figure out which species it could be. We are standing right under the tree, and it's difficult to see his head. We ask the night porter if we can have a look at the bird from a room on the first floor.

Now we can see the owl at eye level. When I look through my telescope, I see an ash-gray owl with two prominent ear tufts in the middle of its head. And then it hits me: It's a Stygian Owl, one of South America's most sought-after owls. We can't believe our eyes; such a rare find, totally unexpected, in the middle of the village, at our very first overnight address. What a flashy start!

With a surface area of 5 million square miles, Brazil is the largest country in South America. About half of the country consists of Amazon rainforest—the entire northwest—and almost the entire east coast consists of Atlantic rainforest. The western central part of the country consists of savannah landscape—the so-called *cerrado*—and the eastern part consists of dry caatinga forest. The vast tropical wetland of the Pantanal is situated in the far west, and the pampa, a subtropical grassy plain, is situated in the far south, at the border with Argentina and Uruguay.

By virtue of this diversity of ecosystems, the country offers an unprecedented wealth of birdlife. There are almost 300 species that cannot be found anywhere else in the world. Many of these species are critically endangered, mainly due to the sheer scale of deforestation in Brazil.

Brazil posed an enormous challenge logistically. Even if I had a whole year, I would never be able to observe all of the country's endemic bird species. Because my visit is just over three weeks long, I had to make necessary concessions. Fortunately, I didn't have to make these choices on my own, because Ies would travel with me during this period.

We made the painful decision to forego much of the northeastern part of Brazil. This part of Brazil is home to some of the most beautiful and endangered species in the world, but due to the enormous distances and the relatively low density of species, the yield would be too low for my Big Year. We decided to include only the southeastern part of the Amazon rainforest near Alta Floresta in the itinerary, as I had already planned to visit this enormously rich ecosystem extensively in Suriname, Peru, and Ecuador. Upon the advice of several Brazilian birders, we would pay a brief visit to the cerrado, the caatinga, and the Pantanal, and further concentrate on the southern part of the Atlantic rainforest, where most of the endemic species occur.

The village of Boa Nova lies against a ridge that forms the border between dry caatinga forest and wet Atlantic rainforest. For this reason, this village is a mecca for birders, as it allows them to observe a large number of species from both ecosystems in the immediate vicinity. Ies is driving, and I am in the passenger seat, leaning all the way forward, full of excitement. In a moment, I will set foot, for the first time, in the Atlantic rainforest, an ecosystem that has fascinated me from childhood.

Just outside the village the dry landscape changes quite abruptly into rainforest. We have arrived at the eastern slope of the ridge. The evaporation of ocean water creates rain clouds, which are blown inland by the easterly wind and then cling onto the ridge. As a result, the eastern slope receives considerably more precipitation than the western slope.

It is clearly wetter here: The road is full of puddles. The drying mud is covered by a carpet of thousands of colorful butterflies. When they feel the vibrations of our footsteps they all fly up in the air, and it is like we are walking through a cloud of confetti.

Almost everything I get to see is new, and by 9 a.m. I've already added more than twenty-five new species. This means I'm officially past 4,000—in less than six and a half months. I would never have thought this possible. I'm starting to feel more and more confident that not only the record but even the magical number of 7,000 species, could be achievable in one year.

Accompanied by a guide from the village, we start our search for the Slender Antbird. This songbird occupies a very specific niche: open caatinga forest with huge ground bromeliads—a flower species closely related to the pineapple—which can be found in transition areas between dry and wet forest.

In Brazil, slash-and-burn cultivation is applied on a large scale. This primitive, destructive farming method consists of burning down a piece of forest in order to transform it into a meadow or farmland. After a number of years, the ground becomes arid and overgrown, at which time the next piece of forest is burned down and the process starts all over again. Due to the relatively nutrient-poor soil in this part of the world, it takes decades for the forest to recover. Usually, nature does not even get that chance. Before it can rebound, the surrounding area has also been cleared and the soil has become completely useless, both for humans and nature.

Antbirds are unable to fly great distances, and, therefore, the agricultural areas breaking up the caatinga forest form impassable barriers. As a result, there is no longer a genetic exchange between the different populations. This process is called genetic erosion, and it makes a population susceptible to inbreeding.

You don't have to be a scientist to see that nature in this area is under great pressure. Much of the original forest has disappeared.

The hills around the village look like a jigsaw puzzle with half of the pieces missing.

We stop at a rather intact stretch of forest at the edge of the Boa Nova National Park. The vegetation is reminiscent of the dry forest of Madagascar. The trees and shrubs are covered with painful-looking thorns, and the forest floor is covered by a thin blanket of dry leaves.

The guide knows exactly where to go. We follow him on a narrow path that ends at a beautiful, immaculate patch of forest. This is clearly the right place, because all around us we see ground bromeliads, sometimes up to 3 feet tall. After a few minutes of waiting, we suddenly hear the characteristic whistling song of the antbird. It's as if someone is inflating an air mattress with a foot pump. Shortly afterward, we see two small, gray-black birds foraging underneath a gigantic bromeliad.

After an exhausting drive of more than 300 miles, we arrive at the Salvador airport in the middle of the night. The two-and-a-half-day round trip to Boa Nova was a huge success, with a total of 117 new species, including many rare endemics.

We try in vain to get a few uncomfortable hours of sleep—in a row of plastic chairs, using our backpacks for pillows—before boarding the plane to Cuiabá, the gateway to the Pantanal.

This is the umpteenth time this year that I have to spend hours at an airport. If there was a world championship in waiting, I would have a chance to win the cup. The ideal waiting time for me is one to one and a half hours; a shorter window of time causes unnecessary stress for me. One unforeseen setback—such as a long queue at the check-in desk, a gate change, or someone ahead of me at luggage control who forgot to take their nail clippers, water bottle, lithium batteries, razors, and a half-empty bottle of wine out of their bag—can trigger an uncontrollable tantrum. In such situations, steam comes out of my ears, and I have to restrain myself from kicking something. Strangely enough, too long a waiting period, such as today, can trigger the same reaction. Once I've written my blog and columns, sorted my photos, and sent all my emails, boredom quickly sets in. Despite being exhausted, sleeping usually proves to be impossible; airports are lit with bright fluorescent lights and there are constant announcements over the intercom. I never dare to put in earplugs, as I'm afraid I won't hear my alarm clock and miss my flight. As I watch time creep by agonizingly slowly, boredom turns into frustration and finally into sheer rage.

The strange thing is that when I'm birding, I don't mind waiting for hours on end. Waiting is, in fact, inextricably linked to the hobby that I love so much. But then I'm usually in a wonderfully beautiful place in the middle of nature instead of in a mind-numbing departure hall. And that makes all the difference.

Cuiabá, Brazil

The Pantanal is a vast wetland area of 68,000 square miles—about twenty times the size of the Everglades. During the rainy season—from December to May—this low-lying, inland delta fills with water from the higher plateaus, turning it into a giant swamp. The water sometimes rises more than 10 feet in a short period of time.

Covering much of southwestern Brazil and a small portion of Paraguay and Bolivia, the area is one of the richest ecosystems in South America. In addition to a huge variety of plant and fish species, it is home to more than 650 species of birds, including the iconic Hyacinth Macaw, the largest parrot in the world. In addition, the area is known for its large numbers of caimans, which are similar to alligators, and capybaras—the world's largest rodents. It is the best place to observe the South American Big Five: the jaguar, giant anteater, lowland tapir, maned wolf, and giant otter.

Our visit to the Pantanal falls in the dry season. During this period the water has largely receded and, as a result, animal life is highly concentrated around pools, lakes, and rivers. Another big advantage is that, this time of year, the area can be reached by car via the Transpantaneira—a 90-mile-long unpaved road with 122 bridges, which crosses the area from north to south.

Just before the exit to the Transpantaneira, our guide, Bianca, suddenly hits the brakes hard.

"Look there, in that bare tree!"

At first, I see only a long blue tail sticking out of a tree cavity, but a little later, the rest of the bird emerges. It is a Hyacinth Macaw. The bird is entirely deep blue in color, except for the bright yellow rear edge of its lower mandible and eye ring. It is more than 3 feet long from the tip of the bill to the tip of the tail. While we are looking at the macaw, a second macaw arrives with raucous noise and lands next to him on a branch. They start preening one another, and then one of the birds disappears into the cavity. This can mean only one thing: They are a couple inspecting a nesting site.

This is an important observation, as Hyacinth Macaws are rare. In total, only 6,500 specimens are left, of which 5,000 live in the Pantanal. They are highly sought after on the black market and can sell for many thousands of dollars each. BirdLife estimates that more than 10,000 macaws were illegally captured and traded in the 1980s. This is done in a horrific way: The nest trees are cut down, and then the young are removed from the nest. This in itself is a real disaster, as macaws reproduce very slowly, but it also destroys the nesting site forever. The young birds are put in tubes with their beaks tied up and smuggled across the border. Many of these animals die of suffocation.

Nowadays, the smuggling of Hyacinth Macaws is subject to astronomical fines and prison sentences. As a result, the trade has declined sharply in recent years. Landowners in the Pantanal are being educated by conservation organizations to ban poachers from their land, and nesting boxes are being put up to provide enough nesting opportunities for these beautiful birds.

⁓—⁓

This time of year can be scorching hot in the Pantanal, but it is uncomfortably cold when we visit. The south of Brazil has been affected by a cold front for a few days, and the temperature is more than ten degrees lower than normal. As a result, many animals are a lot less active, and, hence, they are more difficult to find. We, therefore, decide to mainly focus our attention on the larger, more striking birds. Fortunately there are plenty of them here.

Despite the cold weather, it is business as usual for the waterfowl, and every puddle along the road is full of egrets, Roseate Spoonbills, and stately looking Jabirus, the largest stork species in South America. The banks are dotted with yacare caimans. These cold-blooded crocodilians lie motionless, crisscrossing one another, with their mouths open, waiting for a bit of sun.

We drive south on the Transpantaneira, through a landscape of short grassland and dry patches of forest. We pass a flock of Greater Rheas. Weighing 44 to 60 pounds, this flightless giant is the heaviest bird in the New World. They are reminiscent of the ostriches of Africa and the Emu of Australia, to which they are quite closely related.

I notice that there is a lodge every few miles and that the road is dusty because of the many safari 4×4s. The Pantanal of South America is a bit like the savannas of Africa: an open area with large mammals, crocodiles, and large ratites. The safari 4×4s and lodges complete the picture. This is also the disadvantage of the Pantanal: Hundreds of thousands of tourists come here every year, so it is almost always bustling with activity. On the other hand, this is good news, because if a lot of money comes in, then the Brazilian government will be motivated to protect the area. Fortunately this area is large, about half the size of France, and there is still enough room for the rich wildlife. The real dangers to nature here are the construction of dams and reservoirs in the major rivers that supply the Pantanal, and the pollution of groundwater by factories on the higher plateaus.

The jaguar has decided to hide from us, just like the giant anteater, the lowland tapir, and the maned wolf. We can't complain, however, because during a nighttime ride on the Transpantaneira, an ocelot—a kind of miniature version of the jaguar—crosses the road, and, eventually, we see one of the Big Five: the giant otter. This predator of more than 5 feet long is swimming around in a river right next to the road. Time and again it sticks its head high above the water and produces a high, shrill call. According to Bianca, this behavior means that there is a pup hidden nearby, perhaps under the bridge we are standing on.

Alta Floresta, Brazil

When I look down from the airplane window, I notice the enormous clear-cutting. This used to be a tropical rainforest; now I can see only logging plains. It makes me sick. An area of about ten football fields of Amazon rainforest is cut down every single minute. That is thousands of square miles every year.

We are on our way to Alta Floresta, a town in the northern part of the state of Mato Grosso, which is located in the middle of one of the most vulnerable parts of the Amazon region. Originally, the boundary between rainforest and dry savanna—cerrado—was about 300 miles south of Alta Floresta, but as a result of continuous clear-cutting, that boundary has slowly shifted to the north. From the coastal town of Belém, a 300-mile-wide and more than 1,200-mile-long belt of deforestation now runs all the way to the state of Rondônia on the Bolivian border. This huge area is also known as the Arc of Deforestation.

Unlike many people think, the Amazon rainforest is not a uniform whole. On the contrary, it is immeasurably complex. The Amazon River and its eighteen largest tributaries—sometimes miles wide—act as ecological barriers. Many animal and plant species occur on one side of a river but not on the other side. In fact, this area is a kind of gigantic inland archipelago, where the "islands" are separated from one another by rivers instead of the sea.

The logging industry and the Brazilian agricultural sector are eager to tell the outside world that the Amazon is a big, homogeneous pizza of forest and that if you remove one slice, there will be more than enough left. But you should rather see it as a kind of a super complex quattro stagioni pizza: If you eat one slice, a unique taste will be lost forever.

Due to deforestation, we not only lose precious nature, but the last "free humans" are also in danger of extinction. There are still about a hundred uncontacted tribes in the world, the vast majority of which live in the Amazon region. These are communities that have never had contact with the outside world. While we are at home in our easy chairs in front of the television, there are people who still live in the Stone Age, oblivious to the existence of the modern world. I find this almost inconceivable, comparable to the infinity of the universe or the origin of life on earth. These people remind us of who we really are and where we come from. The real human. Without restrictions from society. Without greed, in perfect harmony with nature.

A few years ago, I saw photos of such an uncontacted tribe, taken from an airplane above the remote border area between Peru and Brazil. These photos show men and women covered in red ocher, firing wooden arrows at an unknown object in the sky. They might have mistaken the plane for an evil spirit that descended on them from heaven to destroy them. And ironically, that is exactly what modern humanity threatens to do to these peoples.

———————

Alta Floresta is located in the middle of one such slice of rainforest, bounded north by the Amazon River, west by the Madeira River, east by the Tapajós River, and south by cerrado. This area is home to a number of bird species that cannot be found anywhere else in the Amazon region. In this area, new species are still being described on a regular basis.

At the airport, we are welcomed by Carlos, the son of the owner of the Rio Azul Jungle Lodge, and Bruno, our guide. Bruno is a Brazilian ornithologist who studies the evolution of birdsong in the Amazon rainforest. He knows the sounds of this area better than anyone else.

The lodge is located along the Azul River, right on the boundary of largely cleared private ranches and a huge, continuous block of Amazon rainforest. Prior to our visit, I looked at satellite images, which clearly showed that, south of the Azul River, there is an increasing number of bare, rectangular areas, and on the other side of the river there is a large patch of green, which extends more than 60 miles to the northwest.

From the ground, the clear-cutting is at least as disheartening as from the plane. Driving north in a taxi, we pass the wooden skeleton of a 130-foot-tall mahogany tree, standing in the middle of a bare felling plain. This is all that remains of the diverse Amazon rainforest that covered this area until a few years ago. Thousands of different animal species used to live here. Now, we can see only a flock of black vultures, perched on top of the lone dead tree, in the blazing afternoon sun.

The bare plains sometimes abruptly change into forest, only to change into felled plains a few miles farther down. Such an isolated patch of rainforest appears intact at first glance, but appearances are deceiving; in reality, only a fraction of the original animal life has remained. The miles-wide logging plains form the same kind of barrier as a large river. Birds and other animals simply cannot make the crossing, and without a fresh supply of life these forest islands will be sterile within ten years.

The forest, in its turn, is completely dependent on animal life: Birds and monkeys eat fruits and defecate the seeds; peccaries—a type of pig—and tapirs keep the undergrowth open; and insects, bats, and birds take care of the pollination of plants. You can think of the forest as a house and the animals as the inhabitants who take care of it. After years of vacancy, the house slowly turns into a ruin.

According to Greenpeace, more than half of the timber coming from Mato Grosso is illegally harvested, while in the state of Pará, this amounts to almost 80 percent. The number of inspectors for a given area, typically several times the size of the Netherlands, can be counted on one hand. Therefore, detecting illegal practices is like looking for a needle in a haystack. Illegally harvested timber is "laundered" and comingled with legally harvested timber by corrupt government officials, who provide it with a CITES or FSC certification (international certificates for responsible forest management) in exchange for bribes. Ranch owners receive astronomical amounts for the hardwood on their land, and the more forest that is cleared, the more livestock they can graze. And what's more, the chance of being caught is very small.

When we arrive at the lodge toward the end of the afternoon, the sun is already low in the sky. We'd spent more than three hours in the car, bouncing over dusty dirt roads. As I stretch my aching limbs, I take in the environment. The lodge, which was built in the middle of a small clearing in the forest, consists of a wooden dining room and two sleeping quarters. Through the trees, I can hear the river flowing, and every few seconds I hear the vibrating, whistling song of a Little Tinamou, a smallish ground-dwelling bird that lives a hidden life on the forest floor of the rainforest.

The Rio Azul Jungle Lodge is a family business, and that is how we are received, as if we were guests of the house. We receive a warm handshake from the owner. His wife hands us each a glass of freshly squeezed passion fruit juice to recover from our long journey. That evening we are served a delicious dinner, which she has prepared herself using local products. When we are enjoying a beer a little later, an Amazonian Pygmy Owl calls from the edge of the forest. I'm already enchanted by this place, even though we haven't even started birding yet.

—————

There are two recently described bird species with a small geographical distribution south of the Amazon River, and the area in the vicinity of the Azul River is the best place to see them: the Tapajós Hermit, described in 2009, and the Bald Parrot, described in 2002. In the morning, we decide to go and look for the parrot, the rarer of the two.

We've taken a boat and bob downstream on the Azul River. There is no wind, and patches of morning fog hang over the river. A morpho butterfly lands on the foredeck of the boat. Rhythmically, it opens and closes its wings, its upper wings glowing fluorescent blue in the first rays of sun. We hear several species of antbirds, trogons, and woodcreepers sing from the forest along the banks, and every now and then groups of Blue-and-yellow and Scarlet Macaws fly over our heads with a lot of noise. The deforestation we saw yesterday seems very far away for a while. Suddenly we see a large, gray-brown mammal with a long snout climbing out of the water. It is a lowland tapir, the largest land mammal in South America. Before disappearing into the forest, it defecates in the water to cover its tracks for any hungry jaguar.

Shortly thereafter, a group of parrots flies toward us, squawking loudly. Bruno shouts, "Bald Parrot!" He has recognized the calls of the birds from afar. As they get closer, we can see the bright red underwings and their striking bare, orange heads. Ies and I are the first Dutch birders to sight this species in the wild.

In order to see the Tapajós Hermit, we have to position ourselves near a row of flowering heliconia plants behind the lodge. This hummingbird species loves the nectar from the heliconia's striking orange-red flowers. It takes a while, but then the little green-and-orange hummingbird suddenly appears in front of us. Its forehead is yellow with pollen, and its wings—beating more than fifty times a second—make a buzzing sound as it flies from flower to flower.

We conclude our stay at the lodge in the way we started: with a boat trip on the Azul River. As the sun slowly sets and we float quietly downstream, Bruno suddenly shoots to his feet.

"Do you hear that?"

We hear a soft "*Ooh*" from the edge of the forest, repeated every few seconds. The sound reminds me a bit of the song of a little bittern.

"That's a Zigzag Heron," Bruno says. The excitement in his voice is clearly audible.

We steer the boat to the side of the river, and Bruno takes out his speaker. It is now getting dark. We remain in absolute silence while he plays the sound. Nothing happens for quite a while, but then Ies sees from the corner of his eye a small, dark shadow flying toward him. He quickly grabs his flashlight. And there it is, in the middle of the beam, sitting on a bare branch: South America's rarest heron.

During our four-day stay at the Rio Azul Jungle Lodge, we see more than 250 bird species, which speaks to the unimaginable biodiversity of this area. Worryingly, just a few miles away, there is a desolate no-man's-land, created by human greed, where life is nearly impossible. The contrast to the Amazon rainforest could not be greater. Will we be able to protect the richest rainforest in the world? Or will the Azul River in the near future be an island of forest, where the songs of the Zigzag Heron and many other unique bird species will fade away over time?

Groningen, Netherlands (2004)

On a normal evening in November, the birding world shook to its foundations. A pair of Pine Grosbeaks had settled in Groningen. Previous Dutch sightings of this finch-like bird all date back to the early twentieth century.

Fortunately, there was still a spot available in Danny's car the next morning. Unfortunately, I had one small problem: I had a math test the next morning, and my teacher, Mrs. Vermeulen, was known for hating truancy. Still, I decided to take the gamble. When we were halfway on our way to Groningen, I called the school to tell them I was sick in bed with fever. While I was on the phone, Vincent and Danny started to imitate the sound of bleating sheep as loud as they could, which didn't exactly add to my credibility.

When we arrived in Groningen, it was just getting light. Dozens of birders had already arrived, all hoping to catch a glimpse of the pair of Pine Grosbeaks. After an hour, we heard shouting going on a bit farther down. Everyone started running like headless chickens; telescopes toppled, and people tripped over fences and curbs. But a little later, we all were truly looking at the Pine Grosbeaks. The male was beautiful pink-red in color while the female was of an inconspicuous moss green.

Usually, these birds are found in the vast boreal forest of North America, Scandinavia, and Siberia, so they rarely come into contact with humans. But the hordes of birders surrounding them didn't seem to bother them. They regarded the hundreds of legs and tripods as a natural part of the landscape. As they rummaged through the flower beds, they could be approached to a yard's distance, and at one point they even hopped between the legs of a photographer.

All that commotion in the otherwise quiet neighborhood of Groningen attracted the attention of the press, and camera crews for NOS, the Dutch public broadcasting network, interviewed the birders and local residents about the Pine Grosbeaks. When I watched the news that evening, I could see myself, clearly recognizable, passing by on national television.

The next day I shuffled into the math classroom as inconspicuously as I could, but just before I was going to sit down, Mrs. Vermeulen called me to her desk.

"So, you were in bed with a fever yesterday?"

While she said this, she pushed her round reading glasses down and looked at me with a stern, determined look.

"Yes, I ate something wrong and I . . ."

She interrupted my stammering and opened her desk drawer. I was utterly surprised when she pulled out a Peterson bird guide. She pushed her reading glasses back up and started flipping through the index. "House Sparrow, Wood Pigeon, Hoopoe . . . I can't seem to find the Pine Grosbeak. Arjan, could you tell me exactly where it is?"

"In the back with the finches, just before the crossbills," I stammered.

"Look, the Pine Grosbeak! It says here that this bird has only been sighted three times in the Netherlands. How sad for you that you were sick in bed, exactly on the day that another couple has been sighted."

I started feverishly picking my brain about how to pull through this.

"Cut the bullshit, Arjan. I saw you on the news yesterday."

However, there was a little smile on her face.

"Next time you want to skip school for something like that, just tell me honestly."

São Paulo, Brazil (2016)

In front of me, I see a man of about thirty-five years of age in baggy skate pants, a New York Yankees cap, and binoculars around his neck. I thought we would be meeting a serious, older man at the airport in São Paulo, not this cheerful punk rocker my own age. Eduardo is one of the best bird guides in the whole of Brazil, and normally he guides only groups from Birdquest, the most exclusive bird travel company in the world.

Eduardo has promised to guide us for the next nine days for a special price, because he thinks my Big Year is "quite a cool adventure." On our way to Itatiaia National Park, he shows us a video on his phone, showing how he lands a backside 360 in a half-pipe with a skateboard.

"I had the choice: Either I became a professional skateboarder or a biologist and professional bird guide. When I broke my ankle in a skating accident, I decided to go for option two."

This has got to be the coolest birder I've ever met.

After a lightning visit to the world-famous Itatiaia National Park, we drive on a winding road to the coast. The landscape along the way is somewhat reminiscent of the foothills of the Alps, but appearances are deceiving. Less than half a century ago, these lovely meadows with grazing cows were part of a vast tropical rainforest.

Historically, the Atlantic Forest stretched from Fortaleza, in the northeast, to Porto Alegre, in the far southeast of Brazil. And inland, it extends to the southeast of Paraguay. All told, it's an area totaling more than a 620,000 square miles. At present, the forest has been reduced to less than 6 percent of what it once was.

Our destination is the village of Pereque, where we will look for the Black-hooded Antwren. This species was discovered and collected in the mid-1800s but was considered extinct not long thereafter. For more than a hundred years, ornithologists searched in vain for this 4-inch bird, until it was rediscovered in 1987 by chance by a passerby who found it in a strip of forest along a public road. And what happened as a result? It was discovered that the habitat of the museum specimen had been incorrectly identified and that the bird had been assigned to the wrong genus (*Myrmotherula*). Antwrens of the *Myrmotherula* genus usually live in treetops. Closer inspection revealed that the Black-hooded Antwren belonged to the *Formicivora* genus, who live close to the ground among dense shrubs and creepers. So scientists had been looking in the wrong place and in the wrong habitat. This rediscovery shows once again that, as a nature conservationist, you should never throw in the towel.

The green mountains merge into the azure blue Atlantic Ocean, with mangrove forests and snow-white beaches as a dividing line. From here, the road winds farther to the southwest. Off the coast, we see small, palm-fringed islands and colorful fishing boats anchored in sheltered, windless bays. When I look up, I see Magnificent Frigatebirds everywhere; these are elegant black seabirds with deeply forked tails and a wingspan of nearly 8 feet.

Just before Pereque, we turn onto a path. We stop at a strip of forest along a small river, a stone's throw from the village. Apart from a few similar strips of forest, the landscape mainly consists of meadows and fields. It is hard to believe that this is the place par excellence to look for the Black-hooded Antwren.

We have to wait a long time, but then Eduardo thinks he can hear the bird a bit farther down. I strain my ears, but I can't hear anything at all.

"Are you really sure?"

We walk to the spot. It is eerily quiet when we get there. Just as I'm getting impatient and want to go back to my old spot, a little bird emerges from the undergrowth. It is pitch black with a contrasting orange brown saddle on the back—a male Black-hooded Antwren!

Our good luck seems to be endless. Every day, we see one or more critically endangered species. This is wonderful on the one hand, but it is also disturbing. Until the arrival of Western humans a few hundred years ago, none of these bird species were endangered, and presumably they had a much larger geographical distribution. So I have mixed feelings about these sightings. The moment we see such a bird, Ies and I are cheering, but when we are back in the car, on our way to the next place, we wonder how humanity could have let it come this far.

The most poignant example is the Marsh Antwren. This species was discovered by scientists only in 2004, in a small reed swamp near São Paulo. A large-scale survey was launched to determine where else this species occurred. The results were extremely disturbing: This antwren occurred in just fifteen tiny wetlands totaling a mere 111 acres. Since then, this bird has disappeared from two of these places due to the construction of a dam. It is threatened by the construction of sand mines and fish farms, as well as by the introduction of an alien grass species, which overgrows and suffocates the swamp areas.

The story of the Black-hooded Antwren and the Marsh Antwren makes me wonder how many animal species from the Atlantic Forest have gone extinct without us even knowing about their existence. I'm afraid there are more of them than we think.

However, I draw hope from the story of eighty-five-year-old Antonio Vicente. As a little boy, he watched his father fell the trees on his land to make way for farming. Not long after he razed the last tree to the ground, the water supply dried up. There were no longer tree roots to hold the precious water. The land became completely unusable.

In 1976, Antonio decided to do something about this. He started replanting native trees on his father's land. His neighbors thought he was crazy. "Where do you think you will get your income in the future?," they asked.

Forty years later, he has planted more than 50,000 trees on the 76-acre plot of land. It has been reconverted into a beautiful rainforest, home to hundreds of animal species. With the trees, the water also returned, giving birth to eight waterfalls, which attract tourists from far and wide. He now runs a successful ecolodge, and many of his neighbors have followed suit.

If one man can achieve this, then mankind as a whole should surely be able to protect the remaining rainforest.

Cruce Caballero, Argentina

A swarm of mosquitoes buzzes around my head, but I'm not allowed to swat them away. Even the slightest movement can be enough to deter the Helmeted Woodpecker. We look through the foliage at a round hole in the trunk of a gnarled old tree. According to Martjan, this could be the sleeping cavity of his elusive object of study.

Martjan is the world's leading woodpecker researcher. As a little boy, his heart was stolen by the Black Woodpecker, and after studying biology at the University of Amsterdam, he has led numerous research projects around the world. He is the authority on the extinct Ivory-billed Woodpecker and Imperial Woodpecker and works for the renowned Cornell Lab of Ornithology. He and his family live half the year in the town of Entre Rios and the other half of the year in the tiny village of San Pedro, a stone's throw from the Argentinian Cruce Caballero Provincial Park, the stronghold of the very rare Helmeted Woodpecker.

It's already getting dark and there's still no sign of the woodpecker. We can hear the hollow song of a Collared Forest Falcon a bit farther down: "*aw-aw-aw-aw-aw-aw-aw wow!*" I look at Martjan, but he gestures that I have to be patient and certainly must not move. Easier said than done, because as the evening wears on, the bloodthirstiness of the mosquitoes seems to increase.

Then we see a bird approaching us with a deeply wavy flight. A nod from Martjan confirms what I already thought: a Helmeted Woodpecker! The bird lands right next to the hole and presses itself vertically against the trunk as only woodpeckers can. It vigilantly looks around, showing off its bright red, fan-like crest feathers. Ies and I take turns looking through the telescope, which reveals every detail of the bird: its finely banded belly, black back and white side neck, and even the brick red color of its iris. The sighting lasts only a few minutes, but it feels like an eternity. When the bird has disappeared into the hole, we silently walk back to the path. We dare to speak again only when we are a few hundred yards away, for fear of disturbing the bird, but then we cannot help but burst out in cheers.

Two days ago, we crossed the border between Brazil and Argentina via the town of Foz do Iguaçu. We only saw the world-famous eponymous waterfalls from the plane, because our tight travel schedule did not allow for a visit. If I have to name one downside to this adventure, it would be that I don't have the opportunity to contemplate the geological, architectural, and cultural wonders of the world. Fortunately, this lack is amply compensated by the overwhelming beauty of nature and the countless, unforgettable encounters with the most amazing animal species.

Brazil was exceptionally favorable to my Big Year, providing me with 554 new species in twenty-two days. My count now stands at 4,527 species, and the world record is quickly approaching. Over the next eight days, Ies and I will traverse northern Argentina with our guide, Juan, from the far northeast through the Iberá Wetlands right through the Gran Chaco—a vast lowland plain of dry forest, which covers large parts of Bolivia, Paraguay, Brazil, and Argentina—to our final destination, San Miguel de Tucumán, a city at the foot of the Andes Mountains. It's a road trip of more than 1,250 miles.

We drive on a dusty dirt road into the Iberá Wetlands. This is the second largest wetland in the world—only the Pantanal is larger—and is Argentina's main water reservoir. This area is of vital importance to swamp birds.

The endless pastures with cows gradually give way to natural grassland. Signs of civilization decrease, and man-made ditches along the road give way to meandering creeks and shallow pools where various heron species prey on fish and frogs. One of the first birds we see is a Maguari Stork, a black-and-white ciconiiform with a strikingly light eye, which looks quite similar to the White Stork found in the Netherlands. As we drive deeper into the area we see more and more Maguari Storks, as well as countless numbers of Snail Kites, an ash-gray, specialized bird-of-prey species with a distinctive hooked beak, that feeds on a diet of freshwater snails.

When evening falls, we settle down on the veranda of a small hotel in the village of Colonia Carlos Pellegrini, on the edge of the Iberá Provincial Reserve, the protected part of the wetland. From here, we have an outlook over the area, which somehow reminds me a bit of the Oostvaardersplassen Nature Reserve in the Netherlands. Here, too, we can hear the calls of different species of rails, but the dense vegetation makes it impossible to see them. However, instead of Bluethroats and reed warblers, we hear brightly colored Scarlet-headed Blackbirds singing from protruding reeds, and instead of our trusted Western Marsh Harrier, here, Cinereous Harriers are hunting above the wet grasslands.

At dinner, we are served meat, as has been the case with every evening meal in Argentina so far. This time it's a giant schnitzel with a few slices of tomato and a portion of white rice. This country must be the ultimate nightmare for vegetarians, as the average Argentinian person consumes more than 240 pounds of meat a year on average. It amazes me that they live to be over sixty years of age with such a diet.

It is precisely this eating pattern, not only in Argentina but around the world, that is causing many bird species of Argentina's wetlands and grasslands to be threatened with extinction. In order to facilitate the meat industry, nature has had to give way. All that livestock farming requires a huge surface area of pasture. According to the World Bank, more than 50 percent of the total land area in Argentina consisted of agricultural land in 2014, while almost the entire west of the country consists of the inhospitable Andes Mountains.

This used to be very different, when cowboys moved their herds across the vast natural grasslands. That's where the high-quality and widely acclaimed Argentinian free-range beef came from. At that time, the consequences for nature were still manageable. The manure produced by these animals even contributed to the fertility of the grassland. Due to the increasing global demand for Argentinian beef, much of this meat now comes from factory farming. Cows are fattened on an industrial scale and 100 to 500 animals at a time are kept most of the year in feedlots as small as 2.5 acres. The enormous amounts of manure and urine that end up in the soil and groundwater contaminate the surrounding nature far and wide. Swaying grasslands with an unprecedented wealth of animal and plant life have been reduced to an ecological desert in which only the most hardened generalists can survive.

The next morning, we leave early to look for the most striking bird species of the Iberá Wetlands, the Strange-tailed Tyrant. There is not a bird in the world that bears any resemblance to this bizarre-looking flycatcher. The males are strikingly black and white, with a broad, light orange beak and a bare, bright red throat. Two long feathers, about one and a half times their body length, protrude from their short tail and flutter behind them like two flags. Most children are obsessed with tigers or giant pandas, but I had a similar fascination for a number of bird species. The Strange-tailed Tyrant was one of them, and I have always dreamed of seeing one in real life.

This species is native to the vast natural wet grasslands of north-eastern Argentina, Paraguay, and Uruguay. But in recent decades, the area of its geographical distribution has become highly fragmented. One of its last strongholds is located here.

We drive at walking pace on a small road that takes us right through the vast grasslands. A flock of Limpkins flies in front of us. These brown, ibis-like birds, with white spots on their necks and wings, are a distant relative of the crane. You can clearly see this in their way of flying. They keep their necks and legs straight and have a shallow, jerky wing stroke.

"There's one!"

Juan hits the brakes hard and reverses as fast as he can. Just a few yards from the car, we see a Strange-tailed Tyrant sitting on barbed wire. Its tail feathers flutter like flags in the wind, and its red throat stands out sharply against its black head and chest.

A bit farther down, we see a female. She is a lot more inconspicuous; with her brown plumage and white eyebrow stripe, she somewhat resembles the Whinchat, which is found in Europe.

In order to save the Strange-tailed Tyrant and numerous other vulnerable species from extinction, nature conservationists created The Grassland Alliance. This conservation initiative talks to farmers in Argentina, Brazil, Uruguay, and Paraguay and tries to convince them to preserve natural grasslands on their land. They have already convinced a number of large landowners, saving tens of square miles of precious habitat. However, the greatest responsibility rests with ourselves, the consumer. If we want to preserve our natural grasslands, wetlands, and forests, we will all have to drastically reduce our consumption of meat.

Tucumán, Argentina

Two Andean Condors are circling high above us. With a wingspan of more than 9 feet, they effortlessly float through the thin air at an altitude of 5 kilometers. Their white collar and patches on their upperwings sharply stand out against their black plumage. The birds are joined by a Variable Hawk. Now, the enormous size of the condors really stands out. The Variable Hawk is by no means a small bird, but it is completely dwarfed by the condors, which are twice as large.

The first images of condors date back from the third millennium BC. These mighty scavengers played an important role in all the great prehistoric Andean cultures, from the Chavín culture to the Nazcas and the Incas. They were associated with the sun god and symbolized health and strength. With this in mind, it is tragic that these birds have sharply declined in numbers, especially in recent years. In Argentina, ranchers try to protect their livestock by depositing poisoned bait for predators such as pumas and foxes, with dire consequences: Dead condors are regularly found next to poisoned sheep carcasses. This has a devastating effect on the population, because condors have at most one young every two years and can otherwise live for decades.

We drive on a winding road through the puna—a type of dry grassland that characterizes the landscape above 3,500 meters of altitude in the central part of the Andes. In the coming days we will travel through this rugged mountain range, situated at the border with Chile and Bolivia.

Giselle and her boyfriend, Facundo, both biology students and birders, sit in the front of the 4×4, and Ies and I sit in the back. Giselle is 6 feet tall and has curly blond hair, which makes her look a bit like Shakira; Facundo, with his round sunglasses and long goatee, would not look out of place at a hippie festival. Although they will have an important exam next week, they decided to take five days off to take us in tow.

While I am slouched in the back of the 4×4, enjoying the impressive landscape, Facundo hands me a maté, the South American alternative to tea; it is made from the dried leaves of the maté plant. In the

157

seventeenth century, when the Jesuits settled in Argentina, tea was still unaffordable—it had to be imported all the way from China, via Europe. When put in hot water, the dried maté leaves turn into a smoky, bittersweet brew that has an uplifting effect, similar to coffee. It is Argentina's national drink and is often served in a wooden gourd (*cuia*) and drunk through a *bomba*, a type of metal straw. Following tradition, I empty the cuia, top it up with hot water, and pass it on to Ies, who repeats this ritual and then gives it to Giselle. In Argentina, this ritual symbolizes conviviality and friendship.

From the grassy puna, the road winds down into an intermontane valley. It rarely or never rains here, and that is clearly visible in the landscape. The vegetation is dominated by cacti and thorny shrubs, and the soil consists of fine sand. Almost every bird we see is new for my list. Every so often we see large flocks of Burrowing Parrots fly by. This parakeet-type of bird is moss-green with a yellow-orange belly and an ash-gray breast. They are perfectly adapted to this arid landscape and nest in colonies in sandstone cliffs, where they use their powerful beaks to excavate zigzag-shaped nest tunnels several yards long, which are connected with each other and form complex labyrinths.

When night falls, we set up our tents along the road among the dry vegetation. While we enjoy the sunset in a folding chair and pass on a maté to one another, I take stock: 56 new species in one day, bringing my total to 4,674 species.

Laguna de Pozuelo, Argentina

"*Mira! Un rhea!*"

Giselle points excitedly into the distance. There, in the middle of the barren salt plain, are three huge ratites with long necks. These are Lesser Rhea.

We are standing at the edge of a vast plain, at an altitude of 3,600 meters. Due to this extreme altitude, there is very little oxygen in the air. Every breath takes effort, and my thinking process is slower than usual; my brain feels foggy. Only my eyes are visible; the rest of my head is wrapped in a cloth to protect my skin from the scorching sun.

It's a few miles' walk to the edge of the lake. The ground under our feet has cracked due to the extreme drought. We see dead birds everywhere. At this altitude the living conditions are harsher than anywhere else in the world; only the strongest can survive.

We can see the lake from afar, but due to the mirages above the salt plain it is impossible to say with certainty where exactly the water begins and the land ends. A pink spot moves between the air vibrations. Is this a mirage? Am I hallucinating because of altitude sickness? Fortunately not. As we get closer, the spot transforms into a flock of several dozen James's Flamingos. They run next to one another through the shallow water, with their necks stretched, while their heads are turning all the time. Half of their beak is bright yellow and its tip is pitch black. The spinning yellow-black spots, combined with their pink plumage, has an almost hypnotic effect.

Two-thirds of all greenhouse gases consists of water vapor. And the warmer the climate, the more water vapor accumulates in the air. At high altitudes, this increase in humidity increases the greenhouse effect. Water vapor consists of tiny water droplets, which act like mini magnifying glasses that intensify the power of the sunlight. For this reason, high-altitude lakes, such as the Laguna de Pozuelo, suffer more from evaporation than those at lower altitudes. In addition, due to this increased solar radiation intensity, the surface water heats up more quickly, causing the water molecules to move farther apart and so the density of the water decreases. As a result, fewer nutrients can seep up from the deeper water. And as a result, there is less and less habitat and less food for the flamingos and many other bird species that depend on these lakes.

A hundred yards away, a sluggish black bird trudges through the sucking mud along the lake shore. He is clearly having a hard time. Thick slabs of mud cling to his legs, and every now and then he sinks down to rest. When he gets up again, his belly feathers are also covered in mud, which only makes his ordeal worse. It is a young Horned Coot, recognizable by its dark beak—adult birds have yellow beaks with three bizarre, brush-shaped, black wattles on top. There is no sign of any of his counterparts as far as the eye can see, and that is a bad sign, because Horned Coots, like all coot species, are social animals. He's at the end of his rope and probably won't survive the night. The very tough conditions at an altitude of 3,600 meters have become too much for him.

Punta Arenas, Chile

As I will visit no fewer than forty countries this year, it is simply impossible to visit all of them during the best time of the year. Sometimes it was necessary to visit certain places around the equator in the middle of the rainy season, because we couldn't choose another itinerary from a logistics point of view. Indonesia, Papua New Guinea, Ghana, and Suriname were just a few examples. We were lucky that El Niño reduced the wetness to some extent. In the far north and the deep south, seasons play a major role in determining when to visit: when it is midwinter in the north, it is midsummer in the southern hemisphere, and vice versa. Therefore, the ideal period for visiting Patagonia is from the end of November to the beginning of March.

When I get off the plane in Punta Arenas, Chile's southernmost city, it's mid-August and the middle of the winter. An icy wind blows right through my thin jacket. The airport is located in the center of the Patagonian steppe, which is covered here and there with snow and ice. In summer months, this endless, rolling plain consists of waving green grass with a carpet of colorful flowers, but now it is rather lifeless. The only birds I see are a few stray Upland Geese.

When my guide, Sebastián, and I drive along the coast the next morning, the horizon slowly turns purple, then pink, and finally an apotheosis of red, orange, and yellow. It seems as if the sky is on fire, and the sunrise is beautifully reflected in the smooth sea with the mountains of Tierra del Fuego as backdrop scenery. I sit in the back seat in silence. Everything is moving very fast this year. I constantly have to rush, and every time I begin to know and appreciate a place, I have to move on. Moments like this are rare and make me realize how incredibly beautiful nature is.

Our destination is a large lake in the Patagonian steppe. According to Sebastián, this is the place for the Magellanic Plover in the summer months. "However, at this time of the year, they move hundreds of miles further north, along the Argentine coast."

I am under no illusion: I will undoubtedly miss this species. A shame, because the Magellanic Plover is unique in many ways. It is a peculiar, light gray wader, with bright pink legs and red eyes. Genetic research has shown that it is so different from other plovers and other waders that it should be assigned a family of its own. Its closest relatives are not, as its name suggests, the plovers, but the Snowy Sheathbill.

A chill wind has come up, and dark clouds are gathering on the horizon. Sebastián looks concerned: "It's going to snow soon, and not just a little bit."

This worries me, as I have a flight to Santiago in a few hours, and in this area it is not unusual that planes cannot take off or land for days on end.

We continue to struggle on against the icy wind. In the distance, I see two light gray birds roaming along the shore of the half-frozen lake. It can't be true, can it? I quickly grab my telescope and look straight into the red eye of a Magellanic Plover. I am perplexed. What are these birds doing here in this no-man's-land? Sebastián was here yesterday, so they must have returned from their wintering area last night. What a godsend. I can watch them while lying on my stomach from just a few feet away, as they peck frantically for food. Every few seconds they pull a worm or insect larva from the freezing mud. They have to consume huge amounts of protein to survive at these temperatures. Will they intuitively sense that a blizzard with wind speeds of more than 50 miles per hour will soon be sweeping over them?

The icy cold has now penetrated my body from the ground up, through my pants and jacket. Shivering with cold, I get ready to walk back to the car. But then Sebastián points to the stony plain behind us. I turn around and see nothing at first, until three stones suddenly start to move. They turn out to be White-bellied Seedsnipes, beautifully camouflaged. These birds usually live high in the mountains, and sometimes birders have to walk for days around windy mountain passes in order to find them. But they apparently winter in the Patagonian steppe, something I was completely unaware of. This is what I love about birding. You always see something totally unexpected. That's exactly why, at home in Amsterdam, I drink my coffee on my balcony, my binoculars at hand. There are always some robins and Great Tits and Blue Tits around, but I have also seen some White-tailed Eagles and even, at one time, a Griffon Vulture flying by.

———————

Soon after, I am on the plane to Santiago. Just in time, as the snowstorm has broken out in full force and, according to the flight attendant, there will be no air traffic to and from Punta Arenas for the next two days. My visit to Patagonia couldn't have been better timed.

Cuzco, Peru (2001)

"Which is the best birding country in the world?"

I didn't have to think long about my answer: "Peru!"

"Then we'll go to Peru this summer," said my father. "It's about time we went on a real father-son trip."

Peru was a regular topic of conversation on The Volcano. According to the few lucky birders who had been there, you could travel there for months and still see new species every day.

This is also the country where the famous ornithologists Ted Parker and Scott Robinson set their legendary Big Day record in 1982. On foot and by dugout canoe, they spotted 331 different bird species around the Cocha Cashu Biological Station, in the Manu Biosphere Reserve, in a span of 24 hours. The current record, 431 species within 24 hours, was set by my good friend Dusan Brinkhuizen, together with three fellow birders, but they had needed to use a car and take a domestic flight to achieve that.

In July 2001, we arrived in Cuzco, the former capital of the Inca Empire, which is situated at an altitude of 3,400 meters. I thought it was the most beautiful city I had ever seen, even though I've always hated cities, even as a child. The thin air rendered the city crystal clear and sunny. On every street corner, there were Quichua, Indigenous highland people of South America, selling their wares in brightly colored costumes. The merchandise ranged from coca leaves to roasted guinea pig and ceviche (a delicious, peppery dish consisting of pickled fish, which gave Peru its culinary fame).

Cuzco was conquered by the Spanish conquistadors in the sixteenth century. In 1572 they executed Túpac Amaru, the last Inca leader. This heralded the definitive end of the Inca era. The Spaniards managed to wipe out this culture in less than a century, killing millions of Incas, which is one of the largest genocides in world history. The conquistadors were aided—albeit unintentionally—by infectious diseases such as the flu, which they brought with them from the Old World.

Nowadays, Cuzco is a modern city, with around half a million inhabitants. However, the Inca culture is still visible in the old center. Some modern buildings are built on top of walls and foundations of the former Inca city. Every single stone has a different shape, yet everything fits together seamlessly. Some of these blocks weigh hundreds of pounds and have as many as thirteen different angles on one side.

Incas didn't use mortar. They worked in complete harmony with the existing contours of the landscape, without using metal tools. This can be clearly seen at Machu Picchu, one of the greatest archaeological finds in modern history, which was directly connected to Cuzco via the world-famous Inca Trail. The exact function of Machu Picchu is lost to time, but archaeologists agree that the site must have played an important, central role within the Inca Empire. The Incas managed to hide it from the conquistadors, which is why it is very well preserved. The walls and houses of Machu Picchu connect almost organically to the rocky ridge on which the city is built. It's still unclear exactly how the Incas accomplished this architectural feat—they may have used sand and harder rocks as abrasives—but one thing's for sure: They were phenomenal engineers, well ahead of their time.

When my father and I arrived at Machu Picchu, it was just getting light. We were the very first visitors that day. We decided we would have the best view of the ruins from Huayna Picchu, the mountain on the north side. We climbed up at a marching pace, and half an hour later we were at the top, covered in sweat. From there we had a magnificent view over the ruins and the surrounding cloud forest, a hundred yards below us.

"We really should go find the Inca Wren now," I said rather hastily. I had the attention span of a seventh-grader with ADHD, and all I could think about was finding that rare bird. At that moment, I couldn't care less about the wonder of the world lying below us.

"Just stay here for a while; take it all in," my father said. "You will thank me later."

And so we sat next to each other in silence, enjoying the breathtaking view, with the height of the Inca civilization literally at our feet. Then the unmistakable song of an Inca Wren broke the silence, and I ran down the mountain like a madman, leaving my father behind in disbelief.

My parents are still amazed that such a philistine eventually went on to study archaeology.

Our journey through Peru came to a wonderful finale: At Pilcopata, my father and I boarded a motorized canoe that took us into the heart of the Manu National Park. For several days we sailed through pristine tropical lowland rainforest and camped on sandbanks along the river. The only signs of human civilization were sporadic Indigenous settlements consisting of a few thatched huts along the riverbank. We saw troops of squirrel monkeys, sometimes a hundred at a time, while macaws and parrots flew over our heads all day long.

On our third morning in the park, as we were sailing leisurely downstream, one of the boat boys suddenly pointed excitedly to the shore: "*Otorongo! Otorongo!*"

I had no idea what he was shouting, but the captain turned the engine off. We stopped the boat and kept it in place by holding a dead branch that protruded above the water. I stared tensely at the shore. At first, all I saw was a wall of greenery above the 10-foot-high bank, but all of a sudden I looked straight into the yellow eyes of a big cat. I thought my heart stopped beating. At a distance of less than 20 yards, a jaguar lay quietly enjoying the morning sun. Right across the bank, its front legs dangled over the edge, and its huge head pointed upward in a stately manner. He reminded me of the Sphinx statues of the ancient Egyptians. The spotting pattern on his fur made him completely disappear into the background. I couldn't understand how the boy had been able to spot him.

We were able to watch the big cat for almost fifteen minutes, until he got up, stretched out—driving his claws into a tree trunk and curling his tail up—and disappeared into the woods.

Iquitos, Peru (2016)

It's sweltering hot, and my shirt is soaking wet just after three steps due to the high humidity. The streets are buzzing with brightly colored mototaxis, the motorized Peruvian counterparts of the Indian rickshaw. The drivers are honking and screaming all the time. The Plaza de Armas is teeming with shoe shiners and merchants selling everything from fake watches to cotton candy and bubble blowers. Atop the steeple of the famous yellow and red St. John the Baptist Cathedral, Black Vultures and Turkey Vultures are sunbathing with their wings spread wide open.

After two long months, Camilla and I finally saw each other again yesterday afternoon at the Lima airport. There was not much time to relax, as our next flight left shortly thereafter.

Iquitos is located in the lowlands of northeastern Peru, in the province of Loreto. It is the jungle city par excellence and can be reached only by boat or plane. Yet, it counts almost half a million inhabitants, making it the largest city in the world that is not connected by road to other cities. Iquitos has grown through the rubber and logging industries. The houses in the old town date back to the early twentieth century, when the rubber trade was in its heyday and foreign traders came to settle here. The city is located on the Amazon

River and therefore, like Manaus, is an important transit port for tropical hardwood.

———————

A little later, we bounce over the water at more than 30 miles per hour in a polyester speedboat. If you've never seen the Amazon with your own eyes, then you need to completely let go of your idea of what a river should look like. The Rhine is little more than a murmuring brook compared to this immense inland sea of water. The Amazon is almost 4,350 miles long, several miles wide in places, and more than 328 feet deep at its deepest point. It has five times more water flowing through it than the second largest river in the world, the African Congo River. For this reason, one fifth of all fresh water entering the oceans from land comes from the Amazon.

The water is murky, as if we are sailing through billions of gallons of milky tea. This color is not caused by pollution but by the large amounts of sediment carried along from the Andes and the sur-rounding rainforest. This does not alter the fact that the river is becoming increasingly polluted due to deforestation—which causes the sediment to be released more quickly—and (illegal) mining. Unfortunately, large amounts of gold are hidden in the Amazon basin. The gold is separated from the gold ore by means of mercury, which then ends up in rivers through forest streams. Indigenous tribes, living along the upper reaches of the Amazon, are the victims of this pollution. In some of these communities, more than 80 per-cent of the people suffer from mercury poisoning. The effects on the human body are irreversible, and symptoms range from lung and kidney failure to brain damage. Ultimately, a slow, painful death is the result. This is what our hunger for prosperity and progress brings along. Maybe you should think about this before you slide a gold ring on your finger.

———————

The Muyuna Amazon Lodge is located along a small tributary amidst pristine *várzea* forest, which floods on a regular basis. That's the rea-son why the entire lodge is built on 6-foot-high piles. At this time of year, we can moor at the jetty and walk the last hundred yards, but during the rainy season, you can move around only by canoe.

Muyuna means "curassow" in the indigenous dialect. This is one of the very few places where birders can see the Wattled Curassow with their own eyes. It is a large black gallinaceous bird with a long tail, curly crest, and unmistakable bright red, rounded bulges at the base of its beak. Less than half a century ago, this species was common and widespread in the western Amazon rainforest, but hunting and habitat destruction have wiped it out almost everywhere. Its spectacular appearance and rarity makes it the holy grail of the Peruvian Amazon rainforest. Ten years ago, I had spent two weeks birding around Iquitos. But at the time, I was broke and could not afford to visit the Muyuna and the Muyuna Amazon Lodge, just 60 miles upriver, which is why I'm even more eager to look for the Muyuna this time.

⌒⌒⌒

If you want to see the Wattled Curassow it's important to get up early, because just after sunrise these birds come down from their roosts to search for food on the ground.

My alarm goes off at half past three. Moises, our guide, is already waiting for us in the dining room with a cup of coffee. "Are you guys ready to search for the Muyuna?"

His intonation bears a striking resemblance to that of Marlon Brando in *The Godfather*. This immediately gives our mission extra-heavy weight. *You come to me and say that you want to see the Muyuna . . .*

We get into a wobbly dugout canoe. It is cloudy and pitch dark. When I stretch my arm forward, I literally can't see the end of it. Fortunately, Moises knows this area like the back of his hand. He sits in the front and does the paddling. The glow of his headlamp reflects in the inky water. Hundreds of moths, attracted by the bright light, fly around his head. We hear the bizarre, ghostly roar of a Great Potoo coming out of the forest, and every now and then, invisible animals splash from the bank into the water: probably black caimans, which hunt piranhas at night.

It is still dark when we dock. Moises ties the canoe to a tree root with a rope. We get out of the boat and climb up the slope.

The first birds are waking up, intuitively feeling that dawn will not be long in coming. As we walk on a narrow forest path, to the left and right we hear the different species of woodcreepers and antbirds starting to sing hesitantly.

Moises points to the ground. "Watch out for bushmasters and fer-de-lances."

It's good that he reminds us that we're walking through a South American rainforest. One moment of inattention can be fatal. If you are bitten by one of these rare venomous snakes in this remote place, the nearest hospital is suddenly an awfully long way away.

It's quickly getting light now. Near the equator, the sunrise and sunset happen a lot faster than in more northern or southern latitudes. Here, the sun rises and sets behind the horizon in a straight line, instead of at an oblique angle. Soon, the first rays of sun pierce the canopy, and just at that moment we hear a deafening roar just above our heads, like a rolling thundercloud. A family of howler monkeys. They look at us from the treetops with suspicious eyes. The roars of the males can be heard miles away, and up this close, they vibrate all the way to your stomach.

Suddenly, we hear a large bird taking to its wings; it is invisible behind the dense foliage. "*That's the Muyuna!*" Moises says excitedly. Judging by its flapping wings, the bird flies deep into the forest. I missed it. I can't help but swear. However, the game apparently has only just begun, and Moises starts to run, slashing overhanging vines and branches with his machete. We run after him. A hundred yards farther, we hear the bird nearby taking off again, and once more we see nothing at all. But then my eye catches a bright red dot. To this day, I still don't understand how on earth I was able to detect it, but the red dot turns out to be the red base of the beak of a male Wattled Curassow, perched a few dozen yards away on a branch, visible only through a tiny hole in the foliage. I put my telescope on my tripod as fast as I can and, miraculously, I have the bird in focus within a few seconds. Despite the great distance, I can take in every detail, zooming in sixty times. Camilla also has the chance to take a look, and then he disappears into the forest, clapping his wings loudly.

Wayqecha, Peru

"*Wuuuuurrrrrrrrrrrrrrrrrrrrrrrrr!*"

It's pitch dark, except for the green light on Miguel's speaker. I hold my breath, but there is only silence around us. I know I have to be patient; night birds usually don't show up easily. We walk past a few more bends. Again he plays the sound, and again it remains silent. It's only on the fifth attempt that we get a reaction, erupting from deep in the forest: a ghostly, rolling scream that makes my hair stand on end. Shortly afterward, a dark shadow flies across the road. In the glow of my flashlight, we can just see a small nightjar with a long, deeply forked tail disappear into the forest. The Swallow-tailed Nightjar.

The past few days we flew from Iquitos, via Lima, to Puno. This city is located on Lake Titicaca, at an altitude of 3,800 meters, in the extreme southeast of Peru. Miguel, our guide for the next three weeks, picked us up from the airport. He had a nice stunt in mind: We would try to see more than a thousand species in a span of three weeks. No one had ever done that before. To achieve this goal, he put together a very ambitious itinerary. In just three weeks he wanted to crisscross all of Peru, starting in Puno and eventually crossing the border with Ecuador in the far northwest. Normally, birders allow at least two to three months for such an itinerary, so doing it in three weeks was a crazy idea. However, according to Miguel, it was not impossible. We are lucky that Juvenal will join us, a very experienced driver and birder, who has worked for several Peruvian bird travel companies almost all his life and knows the country better than anyone else.

After seeing the Titicaca Grebe—a critically endangered flightless grebe species that occurs only on Lake Titicaca—we rushed to the Wayqecha Cloud Forest Biological Station, at the top of Cuzco-Manu Road. This research station is an initiative of the Amazonian Conservation Association (ACA), a Peruvian-American conservation organization. By encouraging local communities to make small changes in the use of their land, they hope to eventually take big steps toward preserving the cloud forests along the eastern slope of the Andes.

In the evening, in the dining room, we meet a young Peruvian biologist who tells us about a pilot project that the ACA started at a local community near Wayqecha. This community has a long tradition of slash-and-burn cultivation.

The earth is warming up more and more as a result of climate change, forcing many plant species that like a cool and wet climate to creep upslope a little bit at a time. In the lower areas it has simply become too hot and too dry, and they will eventually wither away there. Certain orchid and bromeliad species occur on average hundreds of feet higher today than where they were ten years ago. Animals that depend on these plants, such as hummingbirds, insects, and poison dart frogs, are forced to move with these plants. As a result, the lower border of the entire cloud forest ecosystem is shifting farther and farther upward. At the same time, the vegetation along the tree line—at the border of cloud forest and grassland—is being cut down for firewood and to make way for farmland. In summary, there is less and less cloud forest, leaving less and less habitat for the plants and animals that depend on this ecosystem.

The first challenge for the ACA was to find a native, profitable crop that could grow on degraded soil, eliminating the need to burn new patches of forest time and again.

"Do you want to see the solution with your own eyes?"

We follow the young biologist outside. We see a glass greenhouse a bit farther down. She opens the door and proudly says, "Welcome to our plant nursery."

In the greenhouse, there are hundreds of pots with dozens of different types of plants. We walk to a row of purple-blue flowers. She picks up a plant and looks at it lovingly. "This is *tarwi*, the savior of the cloud forest."

Tarwi (*Lupinus mutabilis*) was used by the Incas and has recently been rediscovered as a superfood. The plant is beautiful, grows on arid soil, is more nutritious than soy, and can be used to make flour as well as oil. And there is a market for it in Cuzco, which makes tarwi profitable, as well. The entire community has switched to growing this product, and so they no longer need to cut down the forest.

In addition to impoverishment of the soil, slash-and-burn cultivation also leads to erosion. With the felling of the cloud forest, the tree roots that hold the ground together also disappear. As a result, farmland and sometimes even entire villages are frequently washed away by mudslides. As a countermeasure, a replanting project has been started in the same community.

"In about twenty years' time, there will be another full-fledged cloud forest here," the biologist assures us.

In just a few years, this community has positively changed its entire way of life for themselves and for the benefit of nature.

The ACA has another research station, Villa Carmen (at the bottom of the Manu Road, near the town of Pilcopata), with an experimental farm where all kinds of innovative, sustainable agricultural methods are tested. They are, among other things, experimenting with biochar, a cost-efficient, environmentally friendly alternative to manure, which can be locally produced.

Biochar consists of humus and charred organic material, which is created when plant remains are decomposed at a high temperature, without oxygen. This process sounds more difficult than it is: You cover a smoldering heap of plant remains with earth, and you slowly let it burn. The end product is full of minerals and, when used as a fertilizer, can render even the most arid soil fertile again.

At first, the local farmers around Pilcopata were unwilling to use the biochar on their land—the slash-and-burn tradition was persistent. That's why the ACA set up this experimental farm, so the farmers could see with their own eyes that it worked. The product was provided free of charge and could be picked up from the research station. This turned out to be a success. In just a few years, almost all the farmers have switched to this new way of fertilizing. Most of them now produce their own biochar, which has greatly reduced local deforestation.

A simple solution to a big problem—those are the best kind of ideas, if you ask me. Why is this method not being used more widely? Biochar is cheap and can be produced locally, it curbs deforestation significantly, it releases hardly any CO_2—so the greenhouse effect is also reduced—and the yield of farmland increases, as well.

To finance these types of projects, the ACA relies on donations and money that come in through ecotourism. We pay sixty dollars per person for an overnight stay at Wayqecha, part of which goes to this project. The rest will be used to purchase and protect rainforest and to provide scholarships to Peruvian students. In short, it's a win-win situation for all parties involved. The ACA hopes this model will be adopted by local communities along the entire eastern slope of the Andes. I am absolutely convinced it will work.

Manu Road, Peru

"Wow! That's the most beautiful bird I've ever seen!"

Camilla points excitedly to the Grey-breasted Mountain Toucan that has just landed in a tree right in front of us. This is my 5,000th species this year, and I couldn't have asked for a more beautiful one. It pecks at small green berries, and every now and then it lets out a plaintive, meowing call, which is answered a few seconds later by another Mountain Toucan. We can't see the second bird, as it's hidden behind a blanket of thick fog.

Well, what do you expect in a cloud forest? Most of the time, you literally walk with your head in the clouds. When those clouds are saturated, they return the water to the lowlands in the form of rain. This cycle often repeats itself several times a day. If you fly over the Amazon, you can see this spectacle with your own eyes: Everywhere, columns of water vapor whirl up from the forest like vertical patches of fog. More than half of this water vapor is produced by the trees and plants themselves. They take in water through their roots and release it again in a gaseous state through stomata—microscopic gaps between the cells of the leaves.

The rainforest cannot do without rain and the cloud forest cannot do without fog. As a result, these ecosystems are inextricably linked.

The forest along the higher part of the Manu Road is still largely untouched. However, there is a lot more traffic than the first time I was here. We see many vans and trucks passing by, some transporting tropical hardwood.

In 2001, the settlement of Pilcopata at the bottom of the Manu Road consisted of only a few wooden houses, but now it has become a small town with several thousand inhabitants. The surroundings have been cleared to make way for buildings, rice fields, and banana plantations. Large parts of the forest have been cut down. The roads leading to the Amazonian lowlands act as tentacles for the major cities west of the Andes. Along these roads, the Amazon rainforest is being exploited commercially and gradually being cut down.

We pass through a string of ecosystems, from cloud forest to tropical mountain rainforest, and finally tropical lowland rainforest. The view is impressive: nothing but mist-shrouded, wooded mountain slopes. The Madre de Dios meanders through the bottom of the valley, several miles below us. Beyond the river, we see nothing but greenery: the lowland rainforest of the Manu Biosphere Reserve. More than a thousand bird species have been identified within this single field of view, about as many as in the whole of Europe.

There are birds everywhere. From extensive flocks with various types of colorful tanagers and jays, to rarities like the Chestnut-crested Cotinga. The absolute highlight is our visit to the Andean Cock-of-the-Rock lek, nearby the eponymous Cock of the Rock Lodge.

At the end of a slippery forest path, there is a rickety observation hut. We sit on a wooden bench and do not move an inch. I could already hear their calls from the road: a bizarre growl and meow, more like a pack of cats in heat than the sound of birds. A bright red male comes flying in right away. He starts displaying, right in front of us, for a brown-colored female, sitting a little farther away. When calling, he lets his wings hang down and makes very fast, short little wing strokes, as if he's in a kind of trance. He jumps up and down and occasionally bends over, until he hangs almost vertically down from the branch. Eventually we see about ten males, who all do their utmost to impress the females present. It's an enormous spectacle of color and sound, which reminds me of the courtship of New Guinea's birds-of-paradise.

Santa Eulalia, Peru

We drive in the dark on the Pan-American Highway, the highway that connects the southern tip of South America (the Argentinian city of Ushuaia) with the northern tip of North America (Prudhoe Bay in Alaska). Just after we have left Lima behind us, we turn right and start to drive up a windy road. When it gets light we have already climbed more than a half mile. Now, we are driving on a small road along an almost vertical slope. The ravine to our right is several hundred yards deep. When I stick my head out of the window, I get the scare of my life: I am literally looking straight down. One error by the driver and we're finished. Fortunately, Juvenal has driven up this road dozens of times. When I've recovered from the shock, I have a look again. Far below flows a swirling river, but at this distance it looks no more than a blue ribbon.

This road is known as one of the most dangerous roads in the world, witnessing several fatal accidents every year. Peruvian bus and truck drivers have a tendency for drunk driving, and on a road like this, that's like playing Russian roulette.

———————

Today we are going to try to do the impossible.

The Santa Eulalia Valley is home to three bird species that, due to their rarity, are all on top of the wish list of every birder visiting Peru: the Rufous-breasted Warbling Finch, the White-cheeked Cotinga, and the White-bellied Cinclodes. One should actually set aside two or three full days to visit this area, which would also be sensible with a view to one's health, as the cinclodes occur only above an altitude of 4,600 meters. Miguel has planned to do all this in one day. Driving from sea level to an altitude of almost 5,000 meters in one day, that is madness indeed.

———————

The Rufous-breasted Warbling Finch is an endangered species with a highly fragmented geographical distribution in western Peru. In recent decades, the dry, natural forests in which it lives, have been cleared almost everywhere to give way to the cultivation of exotic eucalyptus, which is used by the local population as a building material. Where the native forest is still intact, the herb-rich soil vegetation—on which the Warbling Finch depends—is disappearing due to overgrazing by livestock. BirdLife estimates the total population of this Warbling Finch at a mere 150 to 700 birds.

We stop on a shrub-covered slope at an altitude of 2,500 meters. There are a few isolated patches of forest to the left and right of the road; for the rest I see only eucalyptus trees and small stepwise-terraced fields along the slope. In recent years it has become increasingly difficult to find the Warbling Finch in this place, according to Miguel. Considering the current situation, I suspect this last remaining suitable breeding habitat will also disappear in the foreseeable future.

As always, I start my search full of optimism, but after three long hours of walking up and down the road in the blazing sun fruitlessly, I slowly start to lose hope. I admire Camilla for going through this ordeal without complaining. She could have preferred to laze around on the beach with a good book, or to have a beer on a terrace, but instead she's here with me in the middle of nowhere, looking for a 4-inch songbird, which, until this morning, she didn't know even existed.

It is a quarter to ten. We have agreed that we have to leave this place by ten at the latest, otherwise our tight travel schedule will be derailed.

We split up to increase our chances. However, I'm about to give up; the sun is baking hot now, and I see fewer and fewer birds. Suddenly, Juvenal comes running toward us, covered in sweat.

"Miguel found one!"

I immediately start to run, and within a few minutes I am standing next to Miguel, gasping for air.

"He disappeared somewhere in the bushes."

That's not what I wanted to hear. The slope he points to is obscured by the thick undergrowth and is really bursting with Rufous-collared Sparrows, whose physique is not much different from that of the Warbling Finch.

For twenty long minutes I try to look into the bushes from the most impossible positions while balancing on the steep slope.

Then a group of sparrows flies up and lands at the top of a bush. Through my binoculars I see that one bird is slightly larger and dark gray in color, with a striking red breast and eyebrow stripe: the Warbling Finch. Yes!

After the bird has flown away, we quickly run back to the car. It's half past ten, so we're more than half an hour behind schedule.

We continue to climb while passing dizzying precipices and breathtaking vistas. Our destination is a *Polylepis* forest at an altitude of 4,000 meters—the home of the White-cheeked Cotinga. Out of the big three, this is the hardest one to find. More than half of birders who visit this area miss this species.

It's noon when we arrive at our next destination, and the sun is burning mercilessly. The only bird activity comes from the ubiquitous Rufous-collared Sparrows; unlike many other birds, this species thrives in degraded habitats. It can be found everywhere humans have left their mark on the landscape, from sea level to altitudes of 4,500 meters. The fact that they are here is not a good sign.

We walk up the slope for several hours, which is not easy at this altitude. The densest pieces of *Polylepis* grow against an impassable, almost perpendicular wall. I look up and strongly suspect that the cotinga will be hidden there somewhere. *So close, but yet so far away.* We have to give up around three o'clock. We didn't have enough to eat and drink, and due to the physical exertion, we are all shaking like a leaf. In the meantime, Juvenal has made some sandwiches and prepared coca tea against altitude sickness. Lunch works miracles, and I slowly get back to myself. While we are clearing the plates and cups, we suddenly hear a soft cackling sound coming from the bushes right next to the road. Miguel jumps up.

"That's the cotinga!"

A little later, a plump songbird with a striped belly and back and striking white cheeks lands atop a bush: a White-cheeked Cotinga. The bird sits quietly and doesn't seem bothered by our presence, which gives us the chance to study it in detail. Usually this shy bird can be spotted only from a great distance through a telescope.

The hairpinned road winds up steeply. It doesn't take long before we've passed the last stretches of *Polylepis*. Now, the landscape consists of only rocks and barren grassland. It's hard to imagine that dozens of bird species manage to live permanently in this rugged, merciless landscape.

The extreme altitude is really taking its toll now. Breathing is getting harder and harder, and I can't quite organize my thoughts.

During my previous visit to Marcapomacocha, ten years ago, I was sick as a dog. I had a pounding headache and had to throw up. When I'd finally gotten to see the cinclodes, it all became too much for me. As I looked through the telescope, tears streamed down my cheeks. But I hardly remember anything about that sighting. I think my brain, due to the lack of oxygen, couldn't process it all that well.

I try to breathe through my nose as often and as deeply as possible—to compensate for the lack of oxygen—and I drink quarts of coca tea. Camilla does the same. It helps. As we cross the pass at 4,700 meters I feel like my head is stuffed with cotton wool, but I don't feel nauseous and, unlike last time, I'm able to sit up straight in the back of the car.

Ahead of us, we see a few small peat bogs among the rocky puna, the habitat of the critically endangered White-bellied Cinclodes. Marcapomacocha is one of the very last places where it still can be found. According to BirdLife's estimate, there are only a few hundred of these birds left. The peat bogs on which they depend are disappearing due to mining activities in the region. The water is used to wash the soil, and the toxic waste products are then dumped into the bogs.

Immediately upon arrival, I see a White-bellied Cinclodes. It is sitting on top of a protruding rock in the peat bog. It is a large songbird with snow-white underparts, a gray hood, and a brown back. With his chest turned in our direction, he is looking out over his territory.

When we get out and carefully walk in his direction, he stays put. He is joined by two more cinclodes, and the three of them start to do some kind of display, their wings unfolded like fans in front of them and their tails straight up in the air. As they frantically dribble back and forth over the rock, they let out a high-pitched series of cries: "*Pipipipipi-WEE-WEE-WEE-WEE!*" After a while, they fly in our direction and start looking for food, right in front of us, only a few yards away. As if we are not there.

Our mission was successful, but I still have a knot in my stomach as we continue our way toward the city of Junín. These birds are so tame and so trusting, even though humans are responsible for their impending demise through our undiminished hunger for natural resources. For this reason, it's also our responsibility to rescue this species. It is important that we learn more about the White-bellied Cinclodes because little is yet known about these birds. What exact requirements do they need for their living environment? At what age are they sexually mature, and how old can they grow? Once we have the answer to these questions, a conservation plan can be put in place to save this beautiful species. We have a long way to go, but with the development of scientific knowledge, it certainly is not impossible.

Bosque Unchog, Peru

Camilla is shivering in her sleeping bag. She has a hat on her head and has put on two layers of clothing, but it still isn't enough to overcome the bitter cold at this altitude. I, to the contrary, had a wonderful night's sleep and was snoring incessantly, to Camilla's frustration. The long winter days I spent as a little boy on the pier of Scheveningen have made me resistant against the cold. We are staying in a tent on the edge of Bosque Unchog, a cloud forest in the mountains of central Peru. Today we will look for the Golden-backed Mountain Tanager, a bird with which I have some unsettled business.

Ten years ago, Vincent, Jelmer, and I walked here for more than ten hours, through the pouring rain, without catching a glimpse of that bird. At that time, Vincent, out of sheer frustration, almost wanted to slap me on the face. That was, he says, because of my "irritating, unbridled optimism." Every half hour I would say, "I think it's getting lighter on the horizon and it'll stop raining soon," while the sky above us remained uniformly dark and it continued to rain relentlessly. In the end, we left Unchog cold, soaked to the bone, and mentally broken, with no Mountain Tanager.

———

At sunrise, I am standing dressed and ready next to the tent. I hear a hollow song coming from the edge of the forest; it's the song of an Undulated Antpitta, a six-second tremolo that rises and then falls again in volume. The first bird I actually see is the Coppery Metaltail, a small, copper-colored hummingbird that inspects the red flowers of a bush one by one in search of nectar. Hummingbirds weigh only a few grams at most and have an extremely fast metabolism, and it takes unimaginable amounts of sugar to keep their tiny bodies going.

I'm ready to go, but when I see Camilla crawl out of the tent, I know I need to temper my enthusiasm. She looks pale and has deep bags under her eyes. It's clear she didn't sleep a wink last night. I try to get her warm, rubbing her with both hands over her chest and back. I give her a bowl of muesli and a cup of hot coca tea. I just hope she will be able to cope with the hardship.

———

The scenery is enchanting: The mist-shrouded cloud forest along the path is full of bamboo, and every square inch of tree trunk or soil is covered with a thick layer of moss. I can almost imagine a troll or a gnome with a knapsack over his shoulder emerging from the forest.

The path leads through moist paramo grassland, up to the edge of the Bosque Unchog valley. There, a large wooden cross marks the point where the path starts to steeply wind down.

The weather is totally different than it was in 2006. The fog has lifted, and a warm morning sun is shining on my face. The warmth of the sun visibly does Camilla good.

Miguel looks at the sky with concern: "The weather is too nice for the Mountain Tanager."

He's right. Too much sun, like too much rain, is detrimental to bird activity. It is, therefore, important to strike early in the morning. But today it doesn't work. In the hours that follow, we walk from one patch of cloud forest to another. Sometimes we wait in silence for half an hour, hoping for a sudden movement in the treetops. However, the forest remains dead silent, and when the clock on my phone reads twelve, we literally haven't seen a bird for over an hour.

"I give up," Camilla says. "I'm going back to the car."

Camilla is absolutely right. I also feel the fatigue, and I've also been so stupid as to not to bring food. It will take Camilla over an hour to walk back uphill, so her shot at the Mountain Tanager is gone. She clearly has a lot less problem with this than I do.

"You're going to miss him now," I say.

"That's okay. Just make sure you see him." She turns around and starts to climb back up.

In the hours that follow, Miguel and I see nothing at all. Slowly, I get disheartened and frustration sets in. I remember reading in Noah's blog last year that he had seen the Mountain Tanager. *He saw it and I didn't,* I think to myself. In a fit of anger, I kick a big rock and hurt my foot.

When the clock strikes four, it's time to walk back to the car. We've been searching continuously for nearly ten hours now. My frustration has given way to despondency. For the second time in my life, I will miss the most beautiful endemic animal of Peru. What a disappointment!

I'm on the brink of tears, and Miguel is also visibly disappointed by this debacle. The way up is tough on an empty stomach. When we get back to the cross, I know my chances are gone. I can't help but swear.

We are only 500 yards away from the car when I suddenly hear a series of high-pitched calls. Shortly thereafter, I see something bright yellow moving through the treetops. I freeze and hold my breath. Then I see the Golden-backed Mountain Tanager emerge—a bright yellow and jet black songbird with a blue cap and subtle reddish-brown spots on its chest. I am delirious with joy, and my despondency and hunger disappear instantly.

Meanwhile, Miguel has had the presence of mind to warn Juvenal through the walkie-talkie. He immediately wakes up Camilla, who had taken an afternoon nap, with the words: "*Mountain Tanager, run!*"

When I look to the left, I see her running toward me. But now the tanagers—there are three—are flying from tree to tree toward the valley. I don't think she'll ever make it. I run toward her and run the last hundred yards with her, pulling her half the way with me. Miguel, meanwhile, keeps a close eye on the birds. Five minutes later we are standing next to him, panting heavily and almost gagging from the exertion.

"They are there." Miguel points to an isolated bush about 20 yards away. This can hardly go wrong. And indeed, soon after, the three Mountain Tanagers appear in all their glory. We have a close look at them before they fly off, deeper into the valley. Camilla arrived just in time. It took more than twenty hours of searching, but my business with this species has finally been settled.

La Balsa, Ecuador

We are finally past the border into Ecuador. We had to wait a few hours because there was a power cut on both the Peru and Ecuador sides of the river. As a result, our passports could not be processed through the computer system. We passed the time with—how could it be otherwise?—birding. This resulted in my last new Peruvian species, the White-shouldered Tanager. This was my 5,412th species this year. Only 707 species to go for a new world record. With still more than three months to go, a lot would have to go wrong to not break that record. But in order to set the bar as high as possible and push myself to the limit, I set my sights on 7,000 species, a number that had hitherto been thought impossible in the birding world.

Peru was the most productive country of my entire Big Year, I can already say that for sure. In twenty-seven days, I recorded no fewer than 1,001 species, 579 of which were new. Especially when you consider that I previously also visited Chile, Argentina, Brazil, and Suriname— where many of the same species occur—it is an incredibly good score.

~ — ~

Juvenal is driving, and Wim sits in the passenger seat, continuously vaping his e-cigarette. He used to be an avid chain smoker until just a few months ago, and the act of putting something in his mouth and inhaling has turned obsessive.

When I visited Peru for the first time in 2001 with my father, Wim was our guide. He had left the Netherlands and moved to Peru a few years earlier, setting up Tanager Tours here. He is now sixty-nine and retired, having handed over this successful bird travel organization to Miguel. However, he would not miss this adventure for the world and decided to travel north from Lima to join our road trip for a few days. (To my great sadness, at the time of the publication of the English version of this book, Wim has passed away. I remember him as a sweet and astute man, an adventurer and a great birder.)

Camilla and I are sitting in the back seat with Miguel between us. We drive through the hills of the far south of Ecuador. I have my laptop on my lap and try to type my daily blog, but that is virtually impossible on this windy road. Our destination is the Tapichalaca Reserve, the only known habitat of the Jocotoco Antpitta.

<hr>

In November 1997, while on an expedition to a remote Tapichalaca Ridge, which borders Peru, the renowned ornithologist Dr. Robert Ridgely and four colleagues heard a bird song unknown to them, a loud "*WOO . . . WOO-WOOhoo!*" After a long search, they found the source of the sound, a spectacular 9.5-inch-tall, gray-brown antpitta, with a black head, white throat, and a unique broad white moustache patch. Immediately they knew they were looking at an undescribed species, for there was not a single bird that even remotely resembled this bird. They named him after Dr. Ridgely himself, *Grallaria ridgelyi*. The local farmers have known the bird for some time as the jocotoco, referring to its distinctive sound. That designation resulted in the English name Jocotoco Antpitta.

Dr. Ridgely and his colleagues quickly determined that the Jocotoco Antpitta was extremely rare, had a very limited geographical distribution, and that its habitat was disappearing at an alarming rate due to deforestation. If things continued like this, the antpitta would be extinct within a few decades of its discovery.

A year later, the Jocotoco Foundation was established, initially to protect the last remaining habitat of the antpitta. Since then, this foundation has purchased fifteen more nature reservations throughout Ecuador, which are home to dozens of endangered and critically endangered bird species. These reservations feature Jocotoco Foundation ecolodges, where the proceeds of ecotourism are used to purchase new land. They work closely with the local population. All the staff and guides are from the surrounding local communities, which has created support for nature conservation and the prevention of deforestation and poaching.

The road slowly climbs up. When we pass the sign with FUNDÀCION JOCOTOCO RESERVA TAPICHALACA, farmland along the road has largely made way for cloud forest. The Jocotoco Foundation carries out reforestation projects on a large scale. They have already planted more than a million trees of up to 130 different native species, filling thousands of acres of forest. We regularly see flocks of Golden-crowned Tanagers, azure birds with a black head and a striking bright yellow crown, looking for food in the tops of the moss-covered trees; they look like brightly colored Christmas ornaments.

In the evening we check in at the lodge. The porch is equipped with sugar-water feeders that are continuously visited by hummingbirds of up to ten different species. In the dining room, the walls are covered with enlarged photos of the local icon: the Jocotoco Antpitta.

Today we are going to cheat a bit. At least that's how it feels to me. In the first years after its discovery, the jocotoco, due to its reclusive lifestyle and rarity, was almost impossible to sight. But a few years ago, researchers at the Jocotoco Foundation started feeding mealworms to a pair of jocotocos every day. The researchers knew the spot just along the path where the birds could be found, as they heard the calls of the male there every day. At first, the birds didn't show up, but when the researchers returned the next day, the mealworms were gone. After months of perseverance, they finally had a hit: The female emerged right in front of them, grabbed a mealworm, and disappeared back into the vegetation. After a while, the male also showed up, reluctantly. Both birds became less and less shy as they associated those weird, two-legged creatures with food.

Today, every birder who visits the lodge is virtually assured of a jocotoco sighting. I have mixed feelings about this. Of course, it's great that you can see such a cool species so easily without disturbing them; however, the challenge is gone. For me, being rewarded with a sighting after suffering for hours in a dense rainforest is the ultimate way to see such a bird. Now, every Yankee-Doodle with enough money can simply come here and put a check in his bird book. However, this year I have to make some necessary concessions, as I am trying to break a world record. So the end justifies the means, and, today I follow the guide like a Boy Scout to the jocotoco feeding station.

After half an hour of walking, Camilla, who is leading the way, suddenly stops. "Look there, ahead of us on the path!"

She steps to one side and less than 5 yards away, a statue-like Jocotoco Antpitta is standing right in front of her in the middle of the path. What a gigantic animal! It is almost the size of a chicken. We are still hundreds of yards away from the feeding station and I was not prepared for this sighting, so I still feel an adrenaline rush. While we're watching the bird, it suddenly starts to call loudly: "*WOO . . . WOO-WOOhoo!*" A little later, the female makes her appearance, and together they hop ahead of us to the feeding station. So they just came to meet us because they were hungry. That's birding in the twenty-first century for you.

El Coca, Ecuador

"Fasten your seatbelts. We're about to land in Puerto Francisco."

The plane makes a sharp turn to the right. When I look down through the window, I see the Napo River meander through the Amazon rainforest. My destination is somewhere down there.

That morning I said goodbye to Camilla in Quito. It was an unforgettable experience to travel with her for a month through the most bird-rich part of the world and to share all those amazing sightings and experiences. From time to time it was a lot of hard work—especially with Bosque Unchog—and I admire Camilla for getting through it gamely and without moaning. Prior to my Big Year, some birders wondered if traveling with my girlfriend would make me lose focus and feared that it would slow me down. I think the opposite has been true, because nowhere have I scored as well as the last five weeks in Peru and Ecuador. Her presence allowed me to motivate myself even more.

Now the time has come for me to go on my own again. The next time we see each other will be at Schiphol in just over three months. Hopefully by then I can call myself the new world record holder.

———

Puerto Francisco de Orellana is also known as El Coca. It is, like Iquitos, a real jungle city. It is sweltering hot, and everything and everyone is drowned out by the honking of mototaxis and the yelling of vendors trying to sell their wares to passersby. El Coca is located at the intersection of the Napo and Coca Rivers and is the starting point for almost every tourist visiting the Ecuadorian Amazon.

On the boulevard I am greeted by Olger, a short man with broad shoulders and the features of an Indigenous South American. My expectations are high. According to Dusan, he is "the best bird guide in the entire Amazon region." And that says a lot, because Dusan is arguably Ecuador's most experienced birder—he's lived in Quito for nearly fifteen years now and is a full-time guide for a well-known birding tour company. Initially, he was going to join me himself, but a few weeks before my visit, things went terribly wrong: During a game of basketball, he stepped on the ball and broke his ankle. This completely ruined our plan. Fortunately, Olger was willing to take over for him.

Olger is a born-and-bred Kichwa from the Sani community. His father taught him to navigate flawlessly through the forest and to hunt in the traditional way—with a blowpipe and darts dipped in poison dart frog toxin. He also learned which plants are edible and which ones have medicinal properties. If I were to get lost in the Amazon rainforest, I would be a dead man; for him it's like walking through a well-stocked supermarket.

———

ARJAN DWARSHUIS

The Sani Lodge is located at the edge of the vast Yasuni National Park. This UNESCO Biosphere Reserve protects nearly 6,200 square miles of rainforest. This is of the utmost importance, as the oil industry has set its sights on the Amazon and doesn't even try to hide this from the public. Just before entering the park, we pass by an oil refinery. The bright yellow flames billowing from the chimneys hurt my eyes. If anything were to go wrong here, the consequences for the environment would be incalculable. In the beautiful documentary *Yasuni Man*, which should be on everyone's list, filmmaker Ryan Patrick Killackey follows an indigenous community for seven years in their battle against the oil companies and the Ecuadorian government, who want to exploit and destroy their way of life and their homeland.

Motorized vehicles are not allowed in the Sani community; they are committed to minimizing their impact on nature and maintaining their traditional way of life. The lodge's revenues are evenly distributed across the community. Their motto is "by the community, for the community." As a result, there is enough money coming in, and so there is no need for hunting or cutting down the forest.

After unloading our belongings from a jetty along the banks of the Napo, we continue by foot on a boardwalk through the várzea forest. The path ends at a small river. The slow-flowing water is the color of Earl Grey tea. That color is created by the tannins released from the decomposing plant remains in the water. We board a dugout canoe and start to paddle. From the dense riparian vegetation, we hear the calls of a Silvered Antbird, a species that is closely linked to this biotope. The trees stand with their roots in the water, while their crowns meet above our heads, like the ceiling of a cathedral.

A little later we sail into a small oxbow lake. If paradise really existed, it would look like this: beds of water lilies floating all around and the sun reflecting on the smooth-as-glass water. Some of the trees are 130 feet tall. Perched atop an overhanging bush, we see a family of Hoatzins, prehistoric-looking birds with spiky crests, russet wings, and bare, blue orbital skin. The young of this primitive, distant relative of the cuckoo have unique claws on their elbow joints that allow them to clamber through riparian vegetation. Furthermore, this is the only bird species that lives almost

186

exclusively on leaves. Normally, birds cannot digest leaves because they cannot ruminate, but the Hoatzin has unique horn-like structures in its goiter that allow it to grind leafy greens before they reach its stomach.

On the farthest bank, there is a traditional wooden building with a thatched roof: the Sani Lodge, my base for the next four nights.

By evening, I'm sitting on the veranda with a beer. I look out over the lake as the sun turns red and slowly sinks behind the trees. Hundreds of thousands of mayflies hover in the air, performing a mesmerizing ballet above the water's surface; they look like gold-colored ballerinas. In the distance, a Crested Owl starts to call out. In that moment, all the stress and pressure surrounding my world record attempt fall from my shoulders. What a heavenly place.

That night, I'm fast asleep when the door of my room opens. I wake up to the soft creaking of the wooden floor and see a dark figure approaching me. In the moonlight shining through the windows, I believe I recognize the contours of an Indigenous man. The man softly shuffles closer, mumbling to himself while he opens my mosquito net and, to my amazement, tries to crawl into my bed.

"Hey, man!" I cry out. "What the hell?"

The man looks startled and almost falls backward, mutters some unintelligible sounds, and then walks out the door again. I'm too tired to think about it any further, so I turn around and go back to sleep.

The next morning I'm not sure if I dreamed the whole incident or if there was actually someone in my room.

"This may sound a little crazy, but I think someone tried to get into my bed last night," I say to Olger as we sip our coffee.

He grins and makes an apologetic gesture.

"That was our shaman. He was on ayahuasca, and he probably confused your room with his. It wouldn't be the first time this happened."

"Wow, that's another nice story to tell at home," I say with a laugh.

Loosely translated from Kitchwa, *ayahuasca* means "creeping plant of the soul." It is a highly hallucinogenic drug extracted from a type of vine and used by indigenous communities in the Amazon basin to interact with ancestors, cure diseases, and make predictions for the future. In the Netherlands and most other Western countries, ayahuasca is on the list of banned substances and is placed in the same category as LSD and magic mushrooms. But for these indigenous communities, this plant has a deep spiritual purpose and plays a very important role in their daily life.

Paz de las Aves, Ecuador

The brothers Rodrigo and Angel Paz grew up in a poor region of Ecuador, two hours from Quito. On their parents' land, they earned merely ten dollars a day. Every morning as they walked from their home to the field, they passed a lekking spot of Andean Cock-of-the-rock, and each time they stopped to admire those beautiful red birds. One day, Angel wondered aloud if there would be any money to be made from this.

And so, in August 2005, the Paz brothers founded the Refugio Paz de las Aves. They built a blind overlooking the lek and put up a sign along the road saying GALLO DE LA PEÑA, Spanish for "cock-of-the-rock." The waiting started. More than two months passed without a single tourist visiting Refugio Paz de las Aves. They almost wanted to throw in the towel when suddenly a birder showed up on their doorstep. They showed the man the cock-of-the-rocks and a few other birds. Afterward, he pressed ten dollars into their hands. By Ecuadorian standards, that was more than a day's salary. Gradually, the place became more famous, and more and more tourists came to Refugio Paz de las Aves to watch the birds.

One morning, Angel was doing some maintenance work on the blind when suddenly a plump, tailless songbird with an orange-brown breast and a heavy beak hopped across the path in front of him. The bird started to ferociously peck at worms. When the bird became aware of Angel, it disappeared into the forest. Immediately thereafter, Angel heard a long drawn out "*hoo-hoo-hoo-hoo-hoo-hoo-hoo-hoo-hoo-hoo-HOO-HOO-HOO-HOO.*" It was a female Giant Antpitta, one of South America's most mythical bird species. He named the bird Maria, after his wife. From that moment on he placed a handful of worms in front of her every day, but each time she turned her back on him. But one day she hesitantly hopped closer, picked up a worm, and disappeared back into the undergrowth. She was the first antpitta in the world known to be hand-fed.

Over time, she was joined by another antpitta, her male, and at one point she even ate from Angel's hand. This news spread like wildfire through the birding world, and Angel was nicknamed "The Antpitta Whisperer." He became the originator of this technique to lure shy forest birds. The Jocotoco Foundation, for example, copied this trick from him.

Currently, he has managed to "tame" no fewer than six different species of antpittas this way, and birders and photographers from all over the world come to Refugio Paz de las Aves to observe the birds. This is fantastic news, as it ensures protection of these rare birds together with their vulnerable habitat, and it raises money for the local population.

There is one drawback to all of this: the Giant Antpitta, like the Jocotoco Antpitta, has lost its mythical status. It reminds me of zoo-like entertainment, despite the fact that the birds are indeed living in the wild. I can still remember the effort I had to go through ten years ago to see the Undulated Antpitta, which is closely related to the Giant Antpitta, in the cloud forests of Peru. I chased him for days on end, until finally I saw him briefly, hopping over a fallen tree trunk. That was quite a different experience!

The night before my visit to Refugio Paz de las Aves, I stay with Dusan in Quito. According to him, all would have to go very wrong to not see the Giant Antpitta tomorrow morning. That very morning the famous antpitta pair had put on quite a show and were photographed by a busload of Japanese nature lovers. The photos on Facebook didn't lie: screen-filling, razor-sharp pictures of Maria eating worms from Angel's hand.

The next morning, I am gently whistling in the car on our way to Refugio Paz de las Aves. The weather is beautiful, and soon I'll meet Maria face to face. And maybe Shakira, the Ochre-breasted Antpitta, will also be prepared to show her sexiest side.

When I arrive, Angel is already waiting for me with a bowl of worms.

"Vamos a ir con Maria?" he asks me.

I follow him. Our first stop is at the cock-of-the-rock lek. The birds put on a great show as Angel stands a few yards away feeding papaya to a family of Dark-backed Wood Quails, a species that used to be notoriously hard to find. Now they literally walk between his legs. A little farther down, a hungry Rufous-breasted Antthrush emerges from the forest edge. He is satisfied with a handful of meal-worms. I add these two normally extremely shy forest birds to my list with somewhat mixed feelings.

Then the big moment has arrived. I take a seat on a bench and Angel puts down a handful of worms. He starts shouting: "*Marieeeeeaaaa, Ma-rieeee-jaaaaaa, venga, venga, venga!*" I wait anxiously, but nothing happens. Half an hour later, Maria has still not appeared. In the meantime, we try our luck at the feeding spot for the Yellow-breasted Antpitta. Fortunately, this one shows up after a few minutes.

We head back to the Giant Antpitta feeding station. "*MAAAAAAA-RIEEEEEE-JAAAAAA?*" Again, nothing happens. I start to get restless. Is this all part of the act? Will she just show up in a minute? But Angel's desperate look speaks volumes: Maria doesn't feel like it today.

It is now after ten o'clock, and still no sign of Maria or her partner. Shakira meanwhile tries to liven things up by sexily moving her behind on a branch. She actually manages to make me smile. How could I not when I see her big brown eyes and her finely striped chest?

We have to give up around noon. Angel shouted his lungs out, but to no avail. I officially missed the Giant Antpitta. I shake my head in disbelief. This sucks. Angel puts his hand on my shoulder to comfort me. A little later, I'm back in the car, returning to Quito, feeling sick at heart. After the Helmet Vanga in Madagascar, this miss hurts the most, especially because I had already considered myself successful in advance. And of course, Noah had managed to see the Giant Antpitta during his Big Year. But things can't always go right, and birds remain birds. As my mother always says, "Arjan, nature cannot be directed."

Montezuma, Colombia

Colombia has a bad reputation with many people, and the country has indeed been a perilous destination for decades. Since the 1960s, it has been torn apart by violent internal conflict. Guerrilla movements such as the FARC, paramilitary right-wing groups, the Colombian mafia (with the ruthless drug lord Pablo Escobar as the main player), and the Colombian military vied for power for many years. Hundreds of thousands of people were killed and nearly 6 million people were forced to flee their homes. People, including politicians and tourists, were murdered and kidnapped for ransom on a daily basis, and large parts of the country were considered no-go zones for a long time.

At the beginning of the twenty-first century, Colombia saw some stability again, as a result of an intensive military campaign and large-scale deployment of police, led by former president Álvaro Uribe. Starting in 2012, ongoing peace negotiations were held between the FARC and the Colombian government, and both sides signed a peace agreement at the end of 2016. The last active guerrilla units have retreated to the Darién Gap, a region at the border with Panama. There is always a chance that violence will flare up again, but Colombians themselves are firmly convinced that the decades-long wave of violence has finally come to an end.

Now that the country has become safer and more accessible, Colombian ornithology has also made massive leaps. Suddenly it became possible for Colombian birders to visit all kinds of areas that, for decades, had been completely inaccessible, such as the Santa Marta and Perijá Mountains. This resulted in a series of spectacular rediscoveries and even a number of newly discovered bird species. Today, Colombia's bird list features nearly 2,000 species with more than 90 endemics, and the country is *the* place to be for every self-respecting global birder.

With the increasing interest from foreign bird-watchers, the infrastructure to facilitate this ecotourism has also improved. Ecolodges have appeared all over the country, and there are now numerous local bird travel companies and guides. Furthermore, there is a large group of young, avid birders and ornithologists who travel to all corners of the country in search of new species and birding spots.

Colombia has quickly developed into one of the great powers of South America when it comes to birding. In fact, it is currently the country with the most species, closely followed by Brazil and Peru.

～一—

We drive uphill in a ramshackle 4×4. The rain has turned the already bad road into a pool of mud. José, our guide, sits in the front. Remco is in the passenger seat, and Gerjon, Garry, and I are crammed together in the back like sardines in a can. I know these three birders from the Netherlands. They are avid world travelers, just like me. When I saw them one morning in October on The Volcano and told them that I wanted to visit Colombia during my Big Year and was looking for travel companions, they immediately offered to come along.

The suspension of the old 4×4 has seen better days. We are shaken about from one side to the other, and with every bump we have to hold on tight so as not to bang our heads hard against the roof. The rain splashes on the windshield and the windshield wipers move frantically back and forth. We're in the cloud forest of Montezuma, in western Colombia, en route to one of the last remaining habitats of the critically endangered Munchique Wood Wren, a 4.5-inch passerine that was only discovered in 2008 due to the former inaccessibility of the area.

There are only 1,500 of these birds left on this mountain ridge and a few more mountains farther south. They only occur in dense dwarf cloud forest at an altitude between 2,250 and 2,640 meters. This very specific habitat is disappearing due to slash-and-burn cultivation; however, this is not the biggest problem for this species. It's a lot more complicated.

The habitat of the Grey-breasted Wood Wren, a bird very similar to the Munchique Wood Wren, is a bit lower on the mountain, in the slightly drier cloud forest. Historically, there is a narrow overlapping zone between the two species around 2,300 meters altitude. Due to clear-cutting in the lowlands and climate change, there is less and less evaporation, and, therefore, less and less cloud formation and precipitation on top of the mountain. This causes the lower limit of the wet dwarf cloud forest, and with it the habitat of the Munchique Wood

Wren, to slowly shift upward. The more dominant Grey-breasted Wood Wren shifts up accordingly, where it expels the Munchique Wood Wren a little bit at a time. This, of course, is a natural evolutionary process, where one species gradually replaces another less well-adapted species. But in this case, the process is triggered by human action, and, therefore, we have to take responsibility.

Finally, we enter the habitat of the Munchique Wood Wren, where the forest is dominated by dense bamboo. In our case, soaking wet dense bamboo, as the rain is still pouring down from the sky. Against our better judgement, we start with our search. On our way up, my folded umbrella served as a shock-absorbing cushion. All its ribs are broken and the stick shows kinks in two places. I hold the crumpled pile of plastic over my head, but it barely stops the rain. A thin trickle of water seeps down my neck, farther down between my shoulder blades, and forms a large wet spot in my underpants, just above my buttocks. I can count myself lucky: Garry's glasses are so fogged up that he can't see a thing. We only have two hours before we have to make our way back down to arrive on time at our next destination.

A little precipitation is fine, but with heavy rain, birds prefer to hide. And that is noticeable. In the first half hour we see nothing at all. However, we frequently do hear a loud, explosive song coming from the dense undergrowth down in the valley. "*Prrrr-prrrr-prrrr-prrrr-prrrr!*"

"That is a yet-undescribed species of tapaculo," José states. "The article will be published in a little over a year. Hopefully you can still include it on your list."

And indeed, as I'm writing this, the Tatama Tapaculo has recently been officially described as a new species. In birding jargon, this is called an "armchair tick," a new species that you can add afterward from the comfort of your home.

By nine o'clock the rain has stopped and a watery sun appears on the horizon. The local bird community comes alive as if by magic to take advantage of this dry moment. Hummingbirds dart across the path in search of nectar-rich flowers, and flocks of tanagers fly from bush to bush.

Suddenly, Remco hears a short, rattling call farther down, which, according to him, sounds exactly the same as the Eurasian Wren found in the Netherlands. We walk up to the dense tangle of bamboo where the sound is coming from, and immediately we see two tiny birds moving through the bamboo—Munchique Wood Wrens. The birds have chestnut-brown upper parts, a striking gray breast, and a thin white eyebrow stripe. Like our Dutch wren, they jauntily stick up their short tails at a perpendicular angle when producing their call.

A little later, we are walking down the road. We hear and see a few more Munchique Wood Wrens. Then we hear the umpteenth rattling call, but this time it is immediately followed by the unmistakable song of a Grey-breasted Wood Wren. The satellite navigation indicates an altitude of 2,350 meters, which is 50 meters higher than the known overlapping zone.

Jardín, Colombia

The cloud-forested mountainsides above Jardín are the stronghold of the endangered Yellow-eared Parrot. This beautiful green parrot species, with its characteristic bright yellow cheeks and forehead, plays the leading role in an inspiring conservation story.

Yellow-eared Parrots once were a common sight in the Andes of Colombia and northern Ecuador. They nested exclusively in hollows of the Quindío wax palm. These parakeets, sometimes in flocks of several thousands, could be found wherever these trees, sometimes reaching a height of 148 feet, towered over the cloud forest. But with the arrival of European immigrants, the parakeets were increasingly hunted, and their habitat was being cleared for farming. In addition, the wax palms were cut down on a large scale every year for Palm Sunday, the Sunday before Easter—the day on which Christian worshippers celebrate the entry of Jesus into Jerusalem. According to tradition, when Jesus made his entry into Jerusalem, he was waved at with palm leaves by bystanders. The predominantly Catholic Colombians relive this moment with leaves of the wax palm. It's a real disaster for a slow-growing tree species, which only blooms for the first time after twenty-five years and can live for up to five hundred years.

At the end of the twentieth century, the Yellow-eared Parrot was considered extinct, but against all expectations, it was rediscovered in 1999 near the village of Roncesvalles by Enter Salaman, a close friend of José. However, they counted a total of merely sixty-one birds, the proverbial last of the Mohicans. Salaman and José noticed that the parakeets were raising hardly any young. There were still a few wax palms between the fields, but the surrounding cloud forest had disappeared. Therein lies the problem: The parakeets depend on old woodpecker nests, because they are unable to carve out nest cavities with their curved beaks. And with the disappearance of the cloud forest, the woodpeckers were gone, too.

Aided by the Loro Parque Foundation, José and Salaman set up an awareness campaign and started an education program. The campaign was successful, thanks in part to the help of a sympathetic local pastor who had a lot of influence on the residents of Roncesvalles. The birds were no longer hunted, and the wax palms were protected. The number of parakeets remained stable and even increased slightly, but the population remained extremely vulnerable; a virus or natural disaster could still wipe out the species. The nesting site was twice caught in the crossfire between the FARC and the military. Miraculously, these incidents had no impact on the parakeet population, but the next time things could turn out a lot worse.

A few months later, another small population of twenty-two Yellow-eared Parrots was unexpectedly discovered in the mountains above Jardín, 200 miles to the north. Here, too, the situation was alarming—the population didn't contain any young birds, and there were hardly any wax palms left. When José and Salaman were watching the Palm Sunday procession in the town square of Jardín, the cause became clear: Everyone was waving palm branches, which meant that several hundred trees had been felled.

It was crystal clear: The felling of wax palms had to be curbed without delay. Here, too, José and Salaman enlisted the help of the local pastor, because it had worked so well in Roncesvalles. It turned out to be a wrong move. According to the pastor, it was God's will that the wax palms were used for Palm Sunday. He convinced the residents that the parakeet story was a lie and instead encouraged them to use the palms.

Fortunately, a lawyer friend told them that in the 1980s the wax palm had been declared Colombia's national tree, so felling these palms was illegal. The law was on their side, but they still failed to defeat the pastor and to prevent the use of the palm leaves. They decided to join forces with other conservationists. Together with eight ornithologists, they founded the ProAves Foundation in the fall of 1999. And so Colombia's first national bird conservation organization came into existence. That same year, ProAves was designated by BirdLife as the official Species Guardian for the Yellow-eared Parrot.

Now that they had the support of a major international conservation organization, an official letter drafted by a judge was sent to the church of Jardín. From that moment on, people who were waving palm leaves on Palm Sunday would be fined. During the next Easter procession, there were already a lot fewer people waving palm leaves, despite fierce protests from the pastor.

In 2003, a new, more progressive pastor was appointed in Jardín. From then on, no one used the leaves of the wax palms anymore. Instead, people started to use the leaves of a local, very common, and fast-growing plant species. The first big win for ProAves was now a fact.

Visiting the parakeets' breeding grounds had been a risky venture, as guerrilla groups were active in the mountains above Jardín. In 2001 a young Dutch ornithologist researching the Yellow-eared Parrot was kidnapped and held hostage for eight months. In 2002, when things became safer under President Uribe, ProAves finally got the chance to engage in active protection work.

They started a cooperation campaign with local landowners. Using funds from BirdLife, ProAves purchased land in the parakeets' nesting area. José and his friends had collected seeds from the droppings of the Yellow-eared Parrots. They used those seeds to replant the barren farmland and coffee plantations around the wax palms. The organic coffee plantations in particular proved to be a mecca for both resident and migratory birds, such as the also endangered Cerulean Warbler. Moreover, the shade of the trees benefited the production and the taste of the coffee. A win-win situation both for humans and birds. In collaboration with the American Bird Conservancy, the production of the first bird-friendly coffee had started.

In 2006, the 465-acre Yellow-eared Parrot Bird Reserve was established. In the absence of suitable natural nesting sites, nest boxes were attached to the palms. This turned out to be a success. Today there are more than 1,500 Yellow-eared Parrots living in the wild again.

ProAves has grown into Colombia's largest conservation organization. There are now twenty-eight ProAves sanctuaries across the country, covering more than 142 square miles and protecting about three-quarters of Colombia's endangered birds. ProAves is a nonprofit organization and uses the same methods as the Jocotoco Foundation. They have built ecolodges at all these reservations, and the money that comes in through ecotourism is used to buy land for establishing new nature reserves. Their goal is to create a network of reservation areas for all the endangered birds of Colombia.

From a high spot, we're looking out over a sloping valley. The sun is low in the sky. Wax palms tower over everything else; they look like green windmills, casting large shadows over the landscape. Almost every palm features a rectangular wooden nest box with a round hole. Some meadows are almost no longer recognizable as such; they are overgrown with lush native vegetation and merge almost organically into the surrounding cloud forest. The song of a Tawny-breasted Tinamou echoes through the valley, and right next to us, a bush with bright yellow flowers is visited by various species of hummingbirds.

José has a blissful smile on his face. "What a beautiful place, isn't it?"

I nod in agreement. While we are enjoying the view, we suddenly hear the unmistakable, nasal sound of a flock of parakeets.

"There they are!"

A flock of several dozen Yellow-eared Parrots comes flying in from the right. Part of the group lands in the crown of a wax palm. Their yellow cheeks glow almost golden in the evening sun. I look at José, who is looking through the telescope, focusing on the birds. Without him, this whole scene would have looked a lot less idyllic.

Scheveningen, Netherlands (1996)

I will always remember my early years of birding; in those days, everything was still new to me. After each morning in the field, I was able to add more bird species to my list, and every single sighting made my heart beat faster. I fantasized about Great Bitterns, Bearded Reedlings, Hawfinches, and other fancy birds that, up to then, I had seen only in my bird book. The anticipation was almost as important as the actual sighting.

I painted the species that I most wanted to see, copying them from the bird book, as if I could bring them to life in doing so. One winter evening, I was painting a Water Rail, a shy reed bird I had never seen before. I lay on my stomach on the carpet in my room, and in utmost concentration I colored the rail's long, curved beak bright red, just like the picture in the bird book that lay open before me. In that same book, I read that this bird makes a sound like a screaming suckling pig.

The next morning my father was already waiting at the kitchen table.

"Why are you up so early, Dad?"

"Do you really think I'm going to let you look for that Water Rail on your own?"

At the crack of dawn, we left on our bikes and headed in the direction of the dunes of Meijendel. A little later, we parked our bikes against the fence along the bike path.

"Those puddles along the path look perfect." I looked at my father with a confident smile. I had started to consider myself an expert on the topic of Water Rails.

We walked slowly, stopping at every puddle. Suddenly the sound of a screaming suckling pig cut through the silence, "*gruIET grroIET gruie gruu!*" I froze. That must be a Water Rail, less than 10 yards away from us, hidden somewhere in the reeds. We sneaked around the puddle and scanned the riparian vegetation with our binoculars after each step. Suddenly, I saw some movement. I nudged my father and we both focused. The reeds moved again. I hardly dared to breathe.

A bright red beak emerged, followed by a piercing red eye.

"See that, Dad? Do you see that?"

My father gave me a squeeze on my shoulder. Together we watched the rail leave its safe haven and walk for a few minutes through the shallow water, completely exposed. His chest and face were ash gray, and his back had the same brown color as the dead reeds. He had a striking pattern of fine vertical black-and-white striping down his flanks, and with every step his jauntily upturned tail moved up and down in a jerky fashion. No matter how many times you look at a bird in a bird book, nothing beats the first time you see it through your own binoculars, taking in every detail.

When I got home I saw that my mother had framed my painting to put it on the wall, as a memento of my first encounter with the Water Rail. Many more encounters would follow, but the first one, in the company of my father, would stay with me forever.

Havana, Cuba (2016)

When I leave South America at the end of October, my count stands at 5,971 species. In almost four months I have sighted 2,325 new species on this continent. Almost everything worked out: The local guides were all excellent, and my traveling companions were fanatics, allowing me to get the most out of it. Moreover, I haven't been ill once, except for a bad cold and a little food poisoning, that is. Unless crazy things happen, the world record will be smashed sometime in the next two weeks. Will 7,000 species be feasible? To get there I will have to see more than 1,000 species in just over two months, and something inside me tells me that that might be too ambitious.

It is the middle of the night when I walk into the arrivals hall of Havana. My father is waiting for me and immediately, with a triumphant look, he presses a pair of cold pizza slices wrapped in napkins into my hands. He took it earlier today from the buffet in his hotel in Varadero, a habit he still has from years of backpacking.

It has always been his dream to see Castro's Socialist Cuba, and now that US economic and political sanctions have been relaxed under President Obama, this may well be his last chance. Very soon, once there are no longer bans on imports from the US, Cuba will look like any other Caribbean island. Cubans do not drive around in old-time cars because they think they are beautiful but because they have no other choice, as it's virtually impossible to buy a modern car due to export sanctions. Almost all of these beautiful old-time cars have cheap Japanese engines, which have been tinkered with dozens of times, and most are on the brink of falling apart. Such an old bone-shaker may prove to be extremely unreliable for a lightning visit during which we have to drive hundreds of miles.

Mysteriously, my father managed to rent a brand new five-door Peugeot.

"I told the owner of the car rental company about your Big Year and got him so excited that he rented his own car to me."

I shake my head in disbelief: Only my father could pull off such a stunt. If necessary, he could arrange for a case of ice-cold beer in the middle of the Sahara.

———

We are the only ones on the road, except for the occasional horse and carriage—most Cubans can't afford a car and a carriage is the best alternative. We drive slowly; the road surface is full of potholes, and the asphalt is cracked from the tropical heat. Our destination is Playa Larga, the gateway to the Zapata Swamp, the best birding spot on the island. Tomorrow, I hope to see my 6,000th species in the presence of my father.

When we arrive at our bed-and-breakfast around two o'clock in the morning, the owner opens the door. Wiping the sleep from his eyes and trying not to yawn, he tells us that Angel Martinez, our guide, will be on the doorstep at six o'clock.

After a short night's sleep, we sit, heavy-eyed, on the terrace in front of our room, sipping at cups of coffee while the sunrise casts an orange glow over the Bay of Pigs. In April 1961, this historic site was the location of a CIA-backed invasion of Cuba. They were attempting to overthrow the socialist regime of Fidel Castro, who had come to power two years earlier. But the attack was nipped in the bud by the Cuban military, and the rest is history.

———

A little later, we drive into the swamp with Angel. The clammy air coming in through the open car windows carries the musty smell of rotting plants. From the impenetrable vegetation along the road, we hear the hollow song of a Key West Quail-Dove. A shiny green bird with a long tail flies off the path. He lands on a branch and turns around, showing his light gray underparts and bright red butt. It is a Cuban Trogon, Cuba's national bird. As soon as we get out to photograph the trogon, we are attacked by countless mosquitoes. The DEET we spray doesn't seem to have any effect. My father's back is covered in dozens of pulsating mosquito bodies, stinging right through his T-shirt.

"You have to be prepared to suffer for supporting your only son in his passion," he says, swatting around wildly.

We drive on. A bit farther down, we see a bird standing in the middle of the road. It is clearly a rail, even though it's hard to see through the wet, slanted windscreen. I hold my breath: Will it be the Zapata Rail? Scientists know virtually nothing about this species' ecology, but they agree on one thing: It is on the brink of extinction. The reason for its decline is very likely predation by the Indian mongoose—a small land predator—and the African catfish, both introduced species. I don't know of a single birder who has ever seen the Zapata Rail. A nest was found only once, in the 1980s, by the American ornithologist James Bond, after whom Ian Fleming, an avid birder himself, named the world-famous British super spy. Yes, James Bond, the coolest man in the world, is a birder, too.

Angel sees my hopeful look and shakes his head. "It's a King Rail."

My father turns the car a quarter turn so that I can see through the open side window. Now I see it clearly through my binoculars: a sturdy brown rail with a striped back and a long, downward curved beak. It is indeed a King Rail, a beautiful species, which is not often seen exposed in the open.

In the afternoon, we visit a wooded area on the other side of the swamp. We are stung to death once again. However, we don't suffer in vain, as Angel shows us some beautiful birds. One of them is the Bee Hummingbird, which is the smallest bird in the world. It is just over 2 inches long (from tail tip to beak tip) and weighs between 1.6 and 2 grams. It buzzes around a butterfly bush like an insect, with more than sixty wing strokes per second.

I take stock: My count now stands at exactly 5,999 species and we still have two hours until it gets dark. Which species will become the magical 6,000th? The Blue-headed Quail-Dove? That would be the ultimate. This endangered pigeon is found only in Cuba and is considered one of the most beautiful pigeon species in the world. In addition, it constitutes an evolutionary enigma: A recent study by the Wilson Ornithological Society shows that the Blue-headed Quail-Dove is not a quail-dove at all and that this species, bizarrely, bears many more similarities to the *Geophaps* pigeon genus, which occurs only in Australia.

We stop at a roadside restaurant. Will we have a drink to celebrate all the beauty we saw today? Angel apparently has a different plan, because we walk past the terrace directly into the kitchen, which borders a small clearing in the woods.

"Wait here."

Moments later, he returns with a handful of rice, which he scatters along the edge of the forest. Almost immediately, two Blue-headed Quail-Doves emerge from the forest. Their deep blue head pattern, with black eye stripe and snow-white moustachial stripe, is phenomenal. These otherwise shy birds are not bothered by our presence and, at a distance of only a few yards, start to peck ferociously at the rice kernels, just like a city pigeon on my balcony in Amsterdam would do. Such a beautiful species, being my 6,000th, in the company of my father . . . Sometimes everything just falls into place.

Cerro Chucantí, Panama

Together with my uncle Fred, who has interrupted a business trip to join me for a few days, and our guides Guido and Nando, we drive east on the Pan-American Highway. After a while we turn into a dusty side road, which we follow until we arrive at a tiny village. We are greeted by an old man with a cigarette in his mouth.

"Ready to meet your best friend for the next five hours?" Guido asks with a grin and then gives the man some instructions in Spanish.

The man disappears around the corner of his house and returns moments later with four scabby horses, which he starts to tack up.

"It's a five-hour ride up to our base camp on the Cerro Chucantí."
I'm petrified at the idea of half a day of riding up a steep slope. The
last time I was on a horse I was merely ten years old. This is going to
be backbreaking.

———————

As a young birder, Guido was always looking for new species. But
after more than ten years of fanatic birding, he had observed almost
all common species around Panama City and in the west of the
country. He knew there were still many new species in the mountainous Darién Gap on the Colombian border, but at that time this
area was far too dangerous to visit because of guerrilla activity. Then
one day in 2003, while studying a topographic map, he discovered an
isolated ridge of nearly a mile long. The ridge, labeled CERRO CHU-
CANTI on the map, was just 75 miles east of Panama City. Would it
be possible that the birds of the Darién Gap were also living there?
He inquired at the university and discovered that there had never
been an ornithological expedition to this area. It had remained a
blind spot on the map.

In 2014 the time had come. Together with a colleague, he left for
the Cerro Chucantí. The climb turned out to be quite a challenge, as
they had to carry dozens of pounds of fieldwork materials and camping equipment. It took them two days to reach their base camp at
700 meters altitude. To their dismay, even in this remote spot, almost
all the forest had already been cleared.

Fortunately, above the base camp was still pristine forest. In the
days that followed, they climbed all the way to the top. Along the way
they encountered a large number of rare animal species, such as the
critically endangered brown-headed spider monkey. His colleague, an
avid botanist, discovered several undescribed plant species. However,
the biggest surprise awaited just below the summit: a Beautiful
Treerunner, a very rare bird species, until then known to occur only in
the Darién Gap.

When he returned to the Cerro Chucantí with some birding
friends a short time later, he saw with great sadness that part of the
forest had been clear cut. The owner of the land had sold it as a logging concession. By their next visit, another couple of acres of forest

had been razed to the ground. In the meantime, Guido discovered even more new animal and plant species nearby at the top of the cloud forest, including tree frogs, salamanders, and even snakes. The isolated ridge turned out to be a hotspot of endemism. He decided to protect this forest at all costs so that these discoveries would not be lost forever.

On the way down they met an old man. They got into a conversation with him, and it turned out that he owned 104 acres of land on the ridge.

"You can have it for $8,200," the man said.

Guido couldn't believe his ears. He had expected an asking price of tens of thousands of dollars. He decided to buy it on the spot. "I'll be back in a few weeks and I'll give you the money."

Guido got the money together, and three weeks later he went back to the Cerro Chucantí. However, he was too late: The man had sold some of the trees on his land to a timber merchant, who had already started cutting them down. Now Guido would have to buy out this merchant, too. A hellish, weeks-long negotiation period followed, but it was fruitful: He could call himself the proud owner of 104 acres of rainforest, full of endemic animal and plant species.

Guido urged friends and birders around the world to invest money in the project so that he could buy up more land. He focused on the plots along the edge of the forest, where the threat of deforestation was most imminent. In 2014, he founded the Adopt a Panama Rainforest Association (ADOPTA). This organization has since bought more than 1,750 acres of forest. His aim is to buy up all the lower forest around the Cerro Chucantí, creating a buffer zone to protect the precious cloud forest at the top.

———～·—～———

"Are you still okay?"

Fred gives me a thumbs up. "I feel like I'm stuck to my saddle, but for the rest I'm fine."

He is a few yards behind me and has a dogged look on his face as his horse climbs up a steep, muddy slope. We've been riding for hours on end and my buttocks and thighs are slowly starting to go numb. The sun is low in the sky and almost disappears behind the hills.

We arrive at the log cabin in the dark. When I get off the horse I almost collapse. I'm already dreading the soreness that undoubtedly will give me a hellish time tomorrow.

Chili simmers on a flickering gas burner. From the trees around the camp, we hear the call of a Choco Screech Owl, which we find after a short while. We admire his magnificent beauty as he sits in the center of the beam of my flashlight. Apart from Guido's Panamanian birder friends, very few people have been here, let alone Dutch people. Just like in the Ekame Lodge in the lowlands of Papua New Guinea, it feels a bit like we've reached the end of the world.

~——~

At three o'clock in the morning, Nando is standing next to my bed. "Are you ready for it?"

When I get out of bed my body feels broken, as if yesterday I had landed under the hooves of a horse instead of sitting on it. After a quick breakfast of cold chili, we start walking.

Nando leads the way. I have my headlamp on, as it still is dark. Because of his unrelenting pace, the blood flow in my legs gets going, and fortunately my soreness disappears quickly.

The path climbs steeply, and every now and then we have to take a break to catch our breath. Two hours later the forest slowly changes into cloud forest: We are now almost at 1,200 meters altitude. I am drenched in sweat and my thighs are burning.

Just after sunrise, we pass the moss-covered wreckage of an American army helicopter. In the 1980s and '90s, helicopters and planes regularly crashed in this area during US military training missions. The fact that the Americans never recovered this wreck shows how remote this place is.

We reach the top around sunrise. From here the path follows the ridge, and walking becomes a lot easier. Almost immediately, we see a group of Tacarcuna Bush Tanagers, a passerine species about which very little is known. These tanagers appear to be common here, because every flock we come across consists mainly of these small, moss-green birds.

When we see the fourth flock, we have a hit. Out of the corner of my eye, I see a chestnut-colored treecreeper-like bird climbing up a mossy trunk. I aim my binoculars and see the unmistakable contrasting white throat. The bird also has a subtle pattern of small, white, raindrop-shaped spots on its belly and neck: a Beautiful Treerunner.

Darién, Panama

The mountains of the Darién Gap form the only interruption of the Pan-American Highway, which is where the "gap" in the region's name comes from. If you look at the map, you can literally see the road ending on either side. In between lies nothing but green, an impenetrable jungle of steep mountain slopes and fast-flowing rivers, where drug smugglers and guerrillas rule supreme. For this reason, the usual way to travel from Panama to Colombia is by plane or by boat over the Caribbean Sea or the Pacific Ocean. Yet there are dozens of people every day who are forced to traverse this green inferno on foot. They are refugees from poverty-stricken countries and war zones in the Middle East, Asia, and Africa. For them, the choice is twofold: either live the rest of their lives in insecurity and abject poverty, or embark on a perilous journey through South, Central, and North America in search of the American Dream.

In South America, their ordeal starts in Brazil or Ecuador, as these countries have lenient immigration laws. In order to finance the flight or boat trip from their country of origin, they often have to sell all their possessions. In the country of arrival, they have to work hard for months under appalling conditions, so that they can pay human traffickers to take them under a false identity across the Amazon rainforest to Colombia. When they finally make it to the Darién Gap, they are completely destitute and often fall into the hands of criminal organizations. They are then smuggled into Panama via a network of forest tracks through this mountainous region. This journey is so long and arduous that many of them never make it to the end. It is the American equivalent of refugees crossing the Mediterranean: a perilous journey in the hope of a new and better life.

In Panama they end up in refugee camps, where they often have to wait for weeks before being taken to Panama City. That's where part three of their journey begins. Without money or identity, they travel right through Central America, where they are likely to be robbed, raped, or killed by Nicaraguan, Guatemalan, or Mexican drug cartels. If they eventually reach the US border, there is a chance that they will be arrested and deported. And then everything will have been for nothing.

The only known place to sight the Dusky-backed Jacamar is situated along the Tuquesa River, where the Pan-American Highway ends and the wilderness of the Darién Gap begins. This place is accessible only by boat.

"Be prepared," Guido says. "There is a refugee camp at the harbor where the boat departs."

A little later we drive into the village. The streets are buzzing with activity. People of African, Asian, and Latin American descent are intermingled.

"Wait here," Guido says, then disappears into the crowd.

We wait at the car. A little farther down, people are queuing up in front of a field hospital, many of them looking sick and emaciated. As I watch them, a dark, very tall man catches my attention. The man looks back at me and smiles kindly. I walk up to him and start a conversation. He speaks fluent English. He says he is from South Sudan and is on his way to the US. His wife suffers from severe migraine headaches, and he is queuing up to buy painkillers with the last of his savings. He used to work as an urban planner, but when civil war broke out in his home country, he had to flee from the violence. He hopes to find a new job in the US so he can provide for his family. I am deeply touched by his story.

Just at that moment, Guido arrives. "The boat has been arranged. We must go now."

I shake the man's hand and wish him all the best.

A little later, we have boarded the boat; I'm on my way to see the Dusky-backed Jacamar. But I can't stop thinking of the man I just met. He is here out of dire necessity. After a hellish journey from

South Sudan, he and his family had to walk from Colombia to Panama. Because he had no choice. And I'm here for my own leisure. Just to watch birds in all their freedom. For some world record that hardly anyone has ever heard of.

A little later I am looking at the jacamar, with mixed feelings. My thoughts are still with that man and his family. The US elections will be held in a few days, and one of Donald Trump's election promises is to close the border with Mexico. Hopefully my South Sudanese friend and his family will reach America in time.

Scheveningen, Netherlands (1996)

We cycled, standing on our pedals, riding up the highest dune top. It was quiet around us. The only other people who ventured outside in the wind and rain on this Sunday morning were a couple of lost cyclists who, like us, were struggling against the pouring rain with grim expressions on their faces. It was the end of November and an autumn storm, which had been raging over the country for the past few days, had blown almost all the leaves from the trees. The beach grass in the foredune had turned into a dull yellow color, and the sea buckthorn bushes along the bike path had been stripped of their orange berries by swarms of hungry thrushes and starlings. Nature was preparing for a long, gray winter.

"What bird are we looking for again?" Maurits asked. A raindrop dangled from his nose, and a wet lock of hair, sticking out from under his hood, almost covered his eyes.

"The Great Bittern," I replied decidedly.

Maurits joined me on almost all my birding adventures, even if he wasn't much of a birder himself. He was with me when I saw my first Eurasian Bullfinch, in the Bosjes van Poot. When my parents rented a holiday house in France, he came with me. Along the way we counted the buzzards and kestrels on the posts along the road. Together we saw a kingfisher for the first time—it was like a fleeting, ice-blue flash over a half-frozen ditch in the Haagse Bos. And when I discovered the nest of a Green Woodpecker in the poplar trees behind my house, he was the first person who came to see them.

And now we were looking for the Great Bittern for the ump-teenth time. So far, this shy marsh bird had managed to hide from me. By now I could even dream of what he would look like: a plump, compact heron, with light green legs, a dagger-shaped beak, and yellow-brown camouflaged plumage. My bird book stated that his song was reminiscent of the sound of a foghorn, and that he took the "pole position" when he felt threatened. In such situations, he would put his beak straight up into the air, where the dark camouflage stripes on his throat and chest would make him completely disappear in between the withered reeds.

When we entered the observation hut in the Kijfhoek, we were numb with cold and soaked to the bone. My old army binoculars were clearly not waterproof, as there was a large black spot on one side every time I looked through them.

The dune lake was largely deserted except for a raft of Tufted Ducks and Common Pochards that were bobbing on the water with their heads tucked in their wing feathers. While Maurits reached for the peanut butter sandwiches my mother had prepared, I peered down the banks. Behind the raft of ducks, a soggy cormorant futilely tried to dry its wings on an overhanging willow branch. A little farther down, a Great Spotted Woodpecker landed on a birch. I focused on a strip of old reed opposite the hut. Slowly my view shifted from left to right. Did I see something moving there? A few seconds later a Blue Tit flew up, right where I had seen the movement. I sighed in disappointment and continued to look. Again I saw a reed move back and forth. "It will be a tit again," I muttered aloud. But suddenly I saw two yellow-orange eyes on either side of the reed. I held my breath and focused a little more, resting the heavy binoculars on the wooden frame. The dark brown reed turned into a dagger-shaped beak and yellow-brown plumage.

"I see . . . a bittern!"

San José, Costa Rica (2016)

"What is a good destination where we can hang out for a few days?" asked Maurits just before my Big Year.

I had to think about it. It shouldn't be too hardcore, like New Guinea or Peru, because Maurits, Pieter, and Harry weren't exactly the most seasoned birders. And, furthermore, Harry has the attention span of a goldfish. It had to be a tropical country, with good infrastructure and a lot of colorful species that were relatively easy to see, and where they could also go out in the evening.

I concluded that Costa Rica would be a good destination.

"You have to prepare properly," I insisted. "And please make sure you all bring binoculars and a bird book."

The alarm goes off at half past three in the morning. My flight was delayed, so I had arrived late last night in San José and got only two hours of sleep. Yet, because of the adrenaline, I am immediately wide awake and jump out of my bed. Could today be the big day? My count stands at 6,087 species, only thirty-two more to go. So it probably will happen today.

"I don't think I've ever gotten up this early," Harry says, plunging into the back seat next to Maurits and Pieter. His bird knowledge is largely limited to his favorite bird, the Short-toed Treecreeper. We almost always see it in the same place in Amsterdam-North, when we traditionally grab a bite to eat on Friday afternoons.

When we arrive at the lodge of El Copal, Juan-Diego and Ernesto are already waiting for us. Pieter Westra, our guide, had announced my visit to the entire Costa Rican birding scene. And it worked, because these guys are some of the best birders in Central America.

We came to this place especially for the Snowcap. This 2.5-inch-long, bright purple hummingbird, with a contrasting snow-white forehead, only occurs in the mountains of Costa Rica and the extreme west of Panama. While the sun hasn't even risen yet, we see a male right in front of us, hanging above the pink flowers of a bush next to the veranda. Because it is still half dark, only his white forehead stands out. It looks like a tiny, snow-white ping-pong ball that is floating through the air.

At sunrise there is an explosion of bird activity. One new species after another passes through the treetops. In addition to the Snowcap, the spectacular White-crested Coquette pays a visit to the irresistible flowers next to the veranda, while we are drinking a cup of coffee at a yard's distance.

"This way, I can see myself becoming a birder," Harry says. "But I don't think I've quite worked out these binoculars yet."

I had to stifle my laughter, as one of his lens caps was still on.

At a tea plantation around the village of Ujarrás, we will look for the Cabanis's Ground Sparrow, a localized bird species found only in the highlands of Costa Rica. Eventually, I add this striking black-and-white sparrow-like bird, with its characteristic rust-brown head cap, as number 6,115. Only four more to go.

Perched atop a bare tree along the road, we see a Grey Hawk, species number 6,116. And a beautiful male Black-crested Coquette flies around the white flowers of a blooming inga tree, species number 6,117. After thirty-three countries and 315 days of nonstop travel, the world record is at my fingertips.

It will have to happen at Ernesto's house. It is located on a coffee plantation where more than 250 different bird species have been identified. Two of those would be new to my list: the White-eared Ground Sparrow and the Buffy-crowned Wood Partridge.

We enter the plantation. My palms feel clammy and my heart is racing. Any bird we see next could potentially be the bird that will allow me to equal Noah's legendary record. My friends also feel that it is getting serious; they are as quiet as cats and look around them with utmost concentration. Then we hear a thin, insect-like call from under a bush.

"White-eared Ground Sparrow!" whispers Juan Diego.

No one dares to move an inch. Left, under the bush, something is moving. I put my binoculars to my eyes and see a dark gray songbird with a striking black-and-white head pattern and a bright yellow neck patch hopping out from under the bush.

Maurits taps me on my shoulder: "Number 6,118!"

After we have examined the sparrow in detail, we sneak farther through the plantation. The Buffy-crowned Wood Partridge is a shy bird, and a sudden movement, or even someone clearing his throat, can betray our presence.

"Let's wait here," Ernesto says softly. "This is the partridge's favorite part of the forest."

I pluck nervously at my binocular strap as we crouch next to one another on the path. The minutes slowly tick by and I start to get cramps in my thighs. Harry has to cough and immediately gets a corrective poke from Pieter. Suddenly a loud, three-tone call echoes through the forest: "*Chu-I-Chuck . . . Chu-I-Chuck . . . Chu-I-Chuck!*" The sound slowly moves away from us.

"Quick, that way!" Juan-Diego gestures for us to follow him.

"Keep a close eye on the ground. They could pass here any moment."

We hold our breath and wait. Out of the corner of my eye I see a movement, right on the slope. I touch Pieter. "There!"

I see two brown-and-yellow partridge-like birds with red legs, a short crest, and a striking pale yellow forehead, walking one after the other through the undergrowth. I only see them briefly, but it's just enough to take in every detail.

I did it! I fall back in the grass and release a deep sigh. I have been looking forward to this moment for years: I am the new world record holder. When I get up, everyone starts cheering. Maurits pulls a bottle of champagne and a box of cigars from under his coat. I uncork the bottle with a loud bang, while Harry shoves a cigar into my mouth and Pieter lights it. I'm a bit too zealous in taking my first sip of champagne and the fizzling foam squirts out my nose. What a great moment, and what a beautiful thing to celebrate with my friends. The record is not only my personal achievement: my friends, family, sponsors, and all guides and fellow birders have had an important part in it. Their confidence and unbridled enthusiasm allowed me to stay totally focused and motivated all along, even when the going got tough.

Chiapas, Mexico

We've been hearing the Nava's Wren singing for minutes now. A simple, repetitive melody: "*Chiu . . . Chu-chiu-chiuu.*" Michael and

Alberto have already seen the bird take off twice, but to my frustration I haven't managed to get it into view and missed it each time.

"There!" Alberto points to the thick vegetation along the road. "About half a yard above that limestone rock, on that horizontal vine."

I shuffle over to him as quietly as possible and peer into the undergrowth with my binoculars. The ground is strewn with limestone rocks and virtually every square inch is covered with lianas. I lose all hope. But suddenly I spot it: a peculiar brown songbird with a short tail, powerful legs, and a disproportionately long, slightly curved beak. As to its physique, it somewhat resembles a kind of miniature Water Rail. With each singing stanza it throws its head back, revealing its white throat. This is species 6,406. And once more, an extraordinarily beautiful one.

I am in Chiapas, in southwestern Mexico, the only region in the world where this wren species occurs. Mexico is the country of wrens: There are no fewer than thirty-two different species, including the largest of all, the Giant Wren.

———————

A few days ago, I arrived in Cancún in the middle of the night. I was picked up from the airport by Michael, the owner of BRANT Nature Tours.

I'd met him six months ago in South Africa. Together with his girlfriend, he joined our road trip for a few days. He told us about his tour company, and since I hadn't yet made any concrete plans for Mexico at that time, he offered to take on that part of my Big Year.

He put together an ambitious itinerary. In sixteen days we would traverse almost every corner of Mexico, a journey of many thousands of miles, with three domestic flights. This way, we could see almost all of Mexico's endemics, except for a few species which only occurred in dangerous areas controlled by drug cartels.

The adventure started in the dry forests of the Yucatán Peninsula. We saw eleven of the thirteen endemics in this area, but oddly enough, we managed to miss the largest and most striking of them all, the Ocellated Turkey. I'll just say that our visit coincided with Thanksgiving, a day when it's better for turkeys to lay low.

On the evening of November 25, we flew to Tuxtla Gutiérrez, the capital of Chiapas, for part two of our trip through Mexico.

We drive on a steep road full of hairpin bends. Far below us, we see the Pacific Ocean, a blue line between the hazy sky and the green coastline. There is not much time to enjoy the breathtaking scenery, because Alberto is in a hurry. It will be dark in two hours, and if we don't see the Giant Wren tonight, we'll have to make a big detour tomorrow morning from the nearest hotel. Due to our tight travel schedule, that's not really an option. According to him, no one has ever seen the Nava's Wren and the Giant Wren in one day. He is determined that we will succeed. But we've got to see the endemic Rose-bellied Bunting first, and the only place that can happen is here, in the La Sepultura Biosphere Reserve.

"Stop! Rose-bellied Bunting!"

Michael has spotted an unmistakable bright blue male sitting in a bush along the road. Alberto hits the brakes hard and the car comes to a halt with squealing tires, just before a hairpin bend. Michael and I get out at the risk of our own lives, after which Alberto maneuvers the car safely into the shoulder. We walk back to the site, and I can breathe a sigh of relief: The bunting is still exactly in the same place and doesn't make any attempt to fly away. Rarely have I seen such a beautiful color combination on a bird. He is an almost unnatural azure blue, with a bright pink belly and a clear, white eye ring.

Alberto points to the sun, which is getting lower and lower in the sky. "We really have to move on now, otherwise we can forget about the Giant Wren."

I take one last look at the bunting, which is now sunbathing, making him look almost like a fluorescent piece of plastic wrap hanging in the tree. Then I get into the car with a heavy heart and we continue our way down. I would have loved to have watched this bird a little longer, but this is part and parcel of this year: Every now and then I have to settle for a frustratingly short sighting. Luckily, I still have the pictures.

We arrive at the coast just before dark. The Giant Wren lives in small family groups and is fairly common in the correct habitat. But are we still in time? The sun is about to set, and we can even hear the call of the first Pacific Screech Owl, a species that, like the wren, can be sighted only here.

Michael grabs his speaker and briefly plays the song of the wren. "Hoping for the best," he says.

But it remains eerily silent. The sun is now only a thin red line in the sky. Just at that moment, we see four birds flying in and landing in a tree, straight above us. They can't possibly be wrens, I think to myself, because these birds are the size of a jay. Our Dutch wren is about 3.7 inches in size. However, immediately thereafter, we hear the liberating song of a Giant Wren. The birds remain there for a few minutes, during which time we can get a good look at their deep chestnut back, black eye stripe, and snow-white underside and eyebrow stripe. Then they fly into a dense, ivy-covered treetop for a well-deserved night's rest. Less than five minutes later we also have a sighting of the Pacific Screech Owl, which is calling in that same tree.

Alberto gives me a triumphant slap on the shoulder. "I knew we would do it."

Mexico City, Mexico

With over 22 million inhabitants, the Mexico City metropolitan area is the most densely populated place in the Western Hemisphere. The metropolis is located at more than 2,000 meters altitude, in the mountain-encircled Valley of Mexico. Nearly 4 million cars spew their exhaust fumes here every day, and most of the time there is hardly any wind. As a result, a suffocating smog is trapped above the city.

It's hard to imagine, but this pinnacle of urbanization once consisted of a network of vast lakes and swamps, with an unprecedented wealth of animal life. Why on earth would you build a city in a swamp? To answer that question, we have to go back 700 years, when an Indigenous tribe from northern Mexico settled on an island in the middle of Lake Texcoco. These few thatched huts would eventually grow into Tenochtitlán, the center of the Aztec Empire. At the beginning of the sixteenth century, this city had more than 200,000 inhabitants, making it one of the largest urban centers of the world at that time (with this in mind, it is almost unimaginable that ancient

Rome, during its heyday in the second and third centuries, perhaps counted as many as a million inhabitants). A legendary monument was the 180-foot-tall, pyramid-shaped stone temple, where human sacrifices were made to the sun god Huitzilopochtli. The opening ceremony of this temple, in 1487, was one of the bloodiest moments in world history: In just four days, about 4,000—according to some anthropologists, as many as 10,000 to 80,000—prisoners of war were sacrificed to the sun god. They were cut open alive by a high priest on a platform atop the temple, their still-beating hearts were removed from their chests, and their lifeless bodies were rolled down the temple steps. The skulls were eventually displayed by the thousands on huge racks and towers, called *tzompantli.*

In 1521, Tenochtitlán was conquered by the Spanish conquistador Hernán Cortés. The city was renamed Mexico City and grew into the capital of New Spain. Due to its location in a valley, the city was regularly plagued by floods. The stagnant waters caused all kinds of diseases, such as malaria and typhus. To prevent this, during the colonial era, the Spanish rulers put thousands of Aztecs to work draining the surrounding swamps and lakes. The landscape of the Valley of Mexico began to slowly change. By the end of the nineteenth century, the city counted about half a million inhabitants. From then on, the population exploded, and Mexico City grew into the metropolis it is today.

Nature has suffered greatly from the centuries-long drainage of the Valley of Mexico. Many bird species that once must have been a common sight are now critically endangered. One species, the Slender-billed Grackle, has long gone extinct. It was last sighted in 1910 in Lerma Swamp, which now accommodates the Olympic stadium, a university complex, and a residential area.

The dangers to nature in this valley are not gone yet; on the contrary, they are increasing. Currently, a new airport is being built, which will take up almost half of Lake Texcoco. Every year, hundreds of thousands of water birds use this place as a wintering area and resting place during their migration to South America.

We leave Mexico City at four o'clock this morning, as the streets quickly turn into a big, honking traffic jam. Sitting next to me is Michael Retter, an American birder who is working on a new field guide covering the birds of Mexico. Michael Hilchey and our local guide, Rafael, sit in the front.

We drive under a gigantic viaduct near the Olympic stadium. The concrete pillars are blackened from the smog, and every square inch is splattered with graffiti. It's hard to imagine that once, Slender-billed Grackles were singing here from swaying reeds with Aztec temples as their backdrop.

Just outside the city, we turn onto a country road, and a little later we drive through a swamp. The morning mist—no doubt mixed with smog—muffles the sound of the city, making it feel like the omnipresent asphalt and concrete are already far behind us.

The first bird we see is an Aztec Rail, which crosses the road just in front of our car. Rafael says that he usually only hears the calls of this shy bird coming from the dense reeds, so we have been very lucky with this sighting. As soon as we get out of the car, we hear the first Black-polled Yellowthroat, and we don't have to look long before we see one. It's a small, olive-colored warbler, with bright yellow underparts and a characteristic black head cap.

We eventually see about ten of them. It is hard to imagine that this species is so endangered. This becomes clear only when you look at a map of its geographical distribution: This species can be found in only a few fragmented wetlands in the Valley of Mexico. Some of them are wedged between Mexico City and Toluca. Toluca has a million inhabitants and will be swallowed up by the Mexico City metropolitan area within a few decades, so by that time, this swamp area, too, will undoubtedly have turned into a concrete jungle.

Nayarit, Mexico

"It will have to happen there." Michael Retter points to the vast pine forest, which is half hidden behind the clouds. "If you manage to see this bird, it will be one of the best species of your entire Big Year."

We are in western Mexico, driving on a windy road into the mountains. Our destination is Rancho La Noria, the place where, just over a year ago, the Cinereous Owl was first documented with photos and sound recordings. Since then, a handful of birders have traveled to this spot to see the owl, but few of them have been successful.

The Cinereous Owl is one of Mexico's most mysterious birds. The first photos and sound recordings of this species were taken only in June 2015, by three American birders. Before that time, the bird was known only from museum specimens, which were collected in the late nineteenth and early twentieth centuries. Furthermore, the identity of this owl was a point of discussion for a long time. Since its appearance is somewhat between that of the Barred Owl of North America and the Fulvous Owl of Central America, it was assumed to be a subspecies of one of these two species. However, based on a genetic study published in 2011 and the conclusion derived from it, it was determined that it should be categorized as a separate species. It only occurs in the mountain forests of Mexico. Before it was "rediscovered," there was no reliable place to see the Cinereous Owl. Most of its assumed geographical distribution is located in drug cartel–controlled areas, where going out at night is no less than a suicide mission.

I always love a nighttime search for an owl. It's the ultimate test of patience and perseverance: the long wait in a dark forest, the tension, the fatigue that starts to play tricks on me over time, the liberating feeling when the bird suddenly starts calling, the inevitable cat-and-mouse game that follows, and the adrenaline rush when the bird is suddenly caught in the beam of light. It encompasses everything. This night's search promises to be one par excellence, as we seek a notoriously rare owl about which little is known.

Michael Hilchey looks concerned at the thin crescent in the pitch-black sky. "We have a new moon, which is bad news. Owls are most active during the full moon."

Michael Retter shakes his head. "That's nonsense. Owls are just most active during a dark night."

My experience, after hundreds of night birding trips, is that when it comes to owls, every birder has their own theories and tactics. One prefers to go out in the evening, the other early in the morning. According to some birders, headlamps are a definite no-no, others swear by them. And so there is a whole series of variables about which opinions are divided. I have only one rule of thumb: You need patience and determination.

There is not a breath of wind, and a mist hangs between the trees. It is winter, and the frogs and crickets are hibernating, making it very quiet in the forest. Every now and then, we think we hear something in the distance. We hold our breath and listen with our hands behind our ears. But every time, it turns out to be a barking dog or a honking car somewhere far below us in the valley.

We hear an owl and are immediately on edge. But then it calls again and we unfortunately have to conclude that it is a Mottled Owl, the widespread, smaller cousin of the Cinereous Owl.

Three hours later, fatigue sets in. My eyes keep falling shut. I have fragments of dreams, from which I invariably wake up with a shock. We decide to get some sleep for a few hours and have another go in the morning.

The alarm goes off just before 4:30. We get dressed and walk back into the forest. It will have to happen now, as when the light breaks through, we will have to move on. Suddenly there is a loud barking right next to us, making us jump; apparently we walked past an invisible fence and alerted a guard dog. Fortunately, the dog stops barking after a few minutes and we continue on our mission.

We walk for a while and then we briefly play the only known sound recording of this owl, which was made exactly at this spot a little over a year ago. We hear a reaction from a great distance, but it is just not sufficient to identify it with certainty.

It will be getting light soon. The first songbirds are already starting their dawn chorus, and I feel that our chances are gone. We've have been looking for almost six hours.

"Let's try one more time, hoping for the best." Michael plays the song one more time: "*Whòhò-Hòhoo-Whoo-hoo-hoohoohoo!*"

Almost immediately I see a gigantic, dark shadow flying toward us, landing straight above us in a pine tree. I aim my flashlight and I see a Cinereous Owl in all its glory. The bird looks straight at us, turning its huge, gray head a quarter turn to the right. What a glorious animal! He has a light facial disc, and his entire body is covered in dark gray-brown stripes; he has a yellow beak and powerful talons with black nails. We are watching him, open-mouthed, and Michael's camera is clicking continuously. Then the owl flies into the forest with silent wing strokes.

When we walk out of the forest it is already light. We really were just in time. We get into the car, exhausted but satisfied. This was indeed an owl mission par excellence.

Sax-Zim Bog, Minnesota, United States

When Ethan and I walk out of the arrivals hall in mid-December, the biting, dry cold slaps us in the face. In front of us, we see a nighttime winter landscape of snow and ice. On the dashboard we read that it is –18°F. What a contrast to yesterday, when we were standing in the subtropical heat, watching a Whooping Crane in a Florida swamp.

Exactly two weeks to go and then my Big Year will be over. I can hardly believe it. Time has gone by so fast. I feel like it was such a short time ago that I started this adventure in Schiphol. I am now thirty-six countries further and my count already stands at 6,671 species.

A week ago I left Mexico and, after a short stopover in Miami, traveled on to Jamaica. I was joined there by Ethan. He had to be in the US for New Year's Eve, and he offered to travel two weeks earlier so that he could join me for a while. This was fantastic news because, knowing Ethan, he wouldn't leave anything to chance. In a span of twelve days, we would visit the states of Florida, Minnesota, New Mexico, and California, and in between we would also make a two-day lightning visit to Jamaica. As in South Africa, Ethan had arranged for us to work with the best local birders and, in many cases, to stay overnight with them, as well.

I have now seen so many bird species in South, Central, and North America that new opportunities are running thin. I have observed almost all of the widespread, more common birds, so only the rare and localized species remain to be discovered. The days with dozens of new species are long behind me, and just like in South Africa—when I had systematically birded the entire eastern side of Africa—the time has come to pick up the crumbs. I have reluctantly accepted that my goal of 7,000 species in one year really is too ambitious.

Fortune smiles on me again: It is crystal clear, windless weather while we drive north through a snow-white landscape. A gigantic snow-storm is predicted to sweep across the northern states of the US starting tomorrow, paralyzing all air traffic for several days. In the media, they are even calling it "Snowmageddon." Fortunately, we will be long gone by then. This year, like a kind of Houdini, I repeatedly have managed to only just avoid all kinds of natural disasters: from floods to tropical hurricanes and earthquakes. I am by no means religious, but you could almost speak of a certain fate.

We are on our way to Duluth, a town at the western tip of Lake Superior. Just north of this is Sax-Zim Bog, a vast moor, with a net-work of fir-tree forests, fields, and lakes. It is a household name among North American birders, as it is the most accessible place to see the typical bird species of the boreal forest. The icon of the area is the Great Grey Owl, one of the largest owl species in the world.

In a roadside restaurant at the edge of Sax-Zim Bog, we meet Sparky, our guide—a huge guy, with a friendly face half hidden behind a shaggy beard.

"Here you go." He presses a thick winter coat into my hands. "You'll need it today."

We get into his car and drive slowly over a frozen country road into the moor. There are endless rows of fir trees to the left and right, and now and then we pass a field with a wooden farmhouse, with smoke curling out of the chimney. Everything is covered in a thick layer of snow. This landscape really has nothing to do with the tropical rainforests I found myself in for most of the year.

In the boreal forest, bird densities are many times lower than in other forests, because there is much less food to be found, especially in the winter. The species spectrum is also smaller, because there are far fewer ecological niches than in, for example, a tropical rainforest. It takes a while before we see the first birds, but then we have a hit: a pair of Ruffed Grouses, which are half hidden at the top of a bush. The bush has lost all its leaves, so their camouflage has no effect. They could just as well have been bright red or purple.

Moments later we see a Northern Shrike—the North American counterpart of the Great Grey Shrike found in Europe—watching the winter landscape closely from the top of a lonely fir tree, looking for prey. His white-gray plumage perfectly matches the snowy world around him, and his black bandit-mask gives him a fearless impression.

We watch birds mainly from the car, because it is bitterly cold in spite of the sunny weather. But when we hear the drumming of a Black-backed Woodpecker, we have to get out. It takes a while before we finally get to see this uncommon species of woodpecker, and he is still frantically pecking at a fir tree. When we get back into the car a little later, my nose and cheeks hurt from the cold.

My main target is without a doubt the Great Grey Owl, even though I already saw this species in Sweden a year earlier, prior to my Big Year. The owl made a deep impression on me, with its spooky gray plumage, its stately appearance, and its expressive, bright yellow eyes. Like the Harpy Eagle, at dusk it has something almost demonic about him, like a forest spirit from another world.

According to Sparky, it will be a bit of a hit-or-miss endeavor in Sax-Zim Bog, because the Great Grey Owl is usually either out hunting right next to the road or somewhere deep in the woods, invisible behind a wall of fir trees.

Toward the end of the morning Sparky gets a phone call. "They have what? Where?" He immediately hangs up and hits the accelerator. "They have a Great Grey Owl! Less than a minute from here."

The tires skid on the icy road. A little later, we arrive at the spot. Three people are standing next to a van: a couple and their local guide, who gives us an apologetic look. The couple is cheering exuberantly, with their hands in the air, and they have written "GGO + 1" in huge letters on the icy rear window of the van. We scan the trees along the road, but the Great Grey Owl is nowhere to be seen.

"He flew off into the forest," the woman says.

If I've learned one thing this year, it's how to deal with setbacks.

We go back to the car and wait in silence for another half hour, but the owl does not come back. Rarely have I been this close to seeing such a cool species without success. Fortunately, the sighting in Sweden is still fresh in my memory.

Pinnacles National Park, California, United States

Ethan and I drive through a typical dry California hilly landscape to Pinnacles National Park. This is the habitat of the largest and most endangered bird in North America, the California Condor. I'm flying to Japan tomorrow morning, so this iconic scavenger might just be the last new species I see in the New World. In the context of my fundraiser for the Preventing Extinctions Program it would be an appropriate species, because no other North American bird has had such a turbulent history of conservation as this one.

The California Condor is a giant black bird with a fierce-looking, bald, pale pink head and contrasting white underwing coverts. It has a wingspan of almost 10 feet and can weigh up to 22 pounds. The Andean Condor, which I saw earlier this year in Argentina, can grow even bigger and weigh up to 33 pounds.

Historically, the California Condor's geographical distribution extended along almost the entire west coast of North America, from the north of Baja California in Mexico to the south of British Columbia in Canada. But by the late 1800s, this area had dwindled. Today, only about 600 of these birds are left. This decline had everything to do with the increase in human population. The condors were shot for their beautiful feathers, their eggs were collected, they fell prey to cyanide traps for coyotes, and they regularly crashed into electricity pylons and other human structures. But the vast majority died from lead poisoning.

Condors are typical scavengers. They feed almost exclusively on the carcasses of large mammals. That is why they have a bald head. When they disappear, head and all, into the abdominal cavity of a dead animal, less rotting flesh and blood gets stuck in their feathers. They track down a meal with their exceptional eyesight, while they soar over the landscape at heights of up to 2.5 miles. Those carcasses used to come from animals that had died of natural causes, but that changed when Western settlers arrived in the eighteenth century. One of the settlers' favorite pastimes was hunting, mainly with lead bullets. When a shot animal managed to get away and eventually died, the bullet was still stuck in its body. When condors that feasted on the cadaver ingested that bullet, the lead was absorbed through their digestive tracts and they died horrific deaths.

The condor population plummeted. By the 1930s, condors no longer existed outside the state of California. In 1953 the state declared the California Condor an officially protected species, and fourteen years later it was assigned endangered species status by the US Fish and Wildlife Service. In the late 1980s, a number of areas were designated as special condor reservations. But it was all to no avail, as hunting with lead bullets still took place on a large scale. In 1985 there were only nine birds left in the wild. At that point, it was decided to undertake the extreme and controversial measure to capture them all in an attempt to save the species and protect the gene pool. The last wild condor, the famous male AC9, was caught in April 1987. From then on, the species was officially extinct in the wild. There were only twenty-two California Condors in the world, and they all lived in captivity.

In the early 1980s, a large-scale breeding program was set up by the US Fish and Wildlife Service, with the participation of several well-known zoos, such as the San Diego Wild Animal Park and the Los Angeles Zoo. This was a huge challenge, because condors become sexually mature only after six years and raise at most one young every two years. But by removing the eggs shortly after laying, the birds were encouraged to lay another one. Those precious eggs were incubated artificially, and the young were raised by hand, making use of a rubber, lifelike condor-head glove in order to prevent the young birds from getting used to humans. The breeding program was a success, and the population grew rapidly. In 1992 the first captive-reared condors were released into the wild. It wasn't until 2002 that a condor was born in the wild again, the first time in eighteen years.

Now several hundred of these majestic birds are living in the wild. In spite of the fact that hunting with lead bullets is banned in California, it still takes place on a large scale. At this point, without the help of the breeding program, the species would still become extinct. This is perhaps the ultimate paradox: The condor cannot live with humans, but it can't live without them now, either.

Pinnacles National Park is characterized by impressive formations of orange rock that are situated in the middle of the park. The condors build their nests on inaccessible ledges along the steepest rock faces. After the reintroduction of the condor back into the wild, this nesting happened for the first time in 2016. The species has been breeding here regularly ever since. The birds are fitted with brightly colored plastic wing tags as well as GPS trackers, which allows researchers to recognize them individually from a great distance and to closely monitor their whereabouts online. It is also useful to trace dead birds and determine the cause of death. With every condor that can be proven to have died of lead poisoning, the argument against hunting with lead bullets gains weight.

Ethan and I are in a parking lot at the bottom of the rock formations. We watch the sky closely, and my telescope is at the ready. We've already seen a number of Turkey Vultures, the much smaller cousin of the condor, but so far we haven't found any trace of the king of California's skies.

Suddenly a bird of prey appears on the horizon. He keeps his wings stiffly spread as he glides effortlessly on the thermals along the rock formation. It's only when a Turkey Vulture flies next to him that his enormous size stands out: It's as if a buzzard is flying next to a Bald Eagle. It is a California Condor, and its 10-foot wingspan casts a long black shadow on the rock wall. I get him in the scope straight away. As I admire the bird, it is joined by another condor. Each has brightly colored wing tags and transmitters on its upper wings. It could very well be that they are a pair and that they are building a nest, because condors lay their eggs in early spring. We can follow them for minutes, until they disappear behind the rock formation. Then we get in the car and drive to Los Angeles. Tomorrow morning I will fly to Vietnam via Tokyo and Taiwan to start the finale of my Big Year.

Da Lat, Vietnam

It is New Year's Day, and a tropical hurricane is raging over Vietnam. It is pouring rain and a gale-force wind bends the treetops. The concrete gutters on either side of the road have turned into swirling mudslides, and the dark rain clouds make it look like night.

"This is the most terrible weather I've experienced all year," I say to Tjeerd, who walks next to me in a soaked poncho.

He looks at me in disbelief. "This is the worst weather you've had in 366 days? We've been walking through the pouring rain for a whole week!"

Tjeerd, Jaap, and André are doing a two-week birding trip through Vietnam and decided to join me on this last day of my Big Year. This is great, of course, because it allows me to end my record year as I had started it: in the presence of friends.

I've been incredibly lucky this year. I can count on one hand the times I've been rained out. Of course I've been out in other heavy downpours, for example, during that terrible day in Madagascar, when I walked all day through a soaked forest and missed the Helmet Vanga. But overall, I've been surprisingly fortunate, even in countries like Papua New Guinea and Ghana, where my visit fell in the middle of the rainy season. I can now also say with certainty that I have not been really sick for one day this year, which is a small miracle. I have no idea how I managed that, but it could very well have to do with willpower. Getting sick was simply not an option. Birding is my greatest passion, and I have enjoyed every minute of this Big Year, even if the odds were against me now and then. I think that this may be the reason that my body tried extra hard to stay healthy for a year, so as not to miss a second of this adventure.

The rain is still pouring down, and our target species, the endemic Grey-crowned Crocias, hasn't appeared yet. We are now drenched to the bone, and the van no longer offers dryness and comfort. The windows are completely fogged up, the benches are wet from our dripping ponchos, and the floor is covered with the mud that came off our mountain boots.

We bravely persevere. At every stop we scan the soaked treetops with our fogged binoculars, hoping for a sudden movement. But it remains eerily silent.

"I only know one last place," our guide says at the end of the morning. "But I haven't seen one there in ages."

Grey-crowned Crocias used to be all over the Da Lat Plateau, but due to a large-scale repopulation project by the Vietnamese government, almost all of that rainforest has been lost to construction projects, roads, and farmland.

We decide to take a chance and drive to the spot: a few acres of degraded rainforest right along the road. It isn't looking very good, especially in this dreary weather. Hoping for the best, we walk on a slippery path into the forest.

Suddenly, we hear the sound of birds in the distance: a flock! Shortly thereafter, we get a sighting of a Vietnamese Cutia and a few Black-headed Parrotbills. In the next treetop, I see a large songbird with a long tail. I step to the side to get an unobstructed view. He has a striking black mask and gray crown, a brown mantle with fine black stripes, and white underparts with thick Song Thrush–like flank stripes: a Grey-crowned Crocias. It turns out that no less than three of these birds are moving along in the flock, and from time to time we get them clearly in view. It took us a morning of suffering, but it certainly was well worth it!

After lunch we drive to another part of the plateau, where we look for another highly localized species: the Black-crowned Fulvetta. It's not looking good. The river has overflowed its banks in many places, engulfing fields and coffee plantations. The plan was to look for the fulvetta in the woods, but the trails have become impassable. We try to wade barefoot through a stream to reach the forest edge, but the fast-flowing water prevents us from doing so safely. We have no choice but to birdwatch alongside the road. On the way, we meet a Vietnamese family who got their car stuck in the mud. With united forces, we manage to get their car out of the mud, but when I look at my mobile phone I see that it is already nearing five o'clock. It will be dark in less than half an hour, and then my Big Year will be over for good.

The Black-crowned Fulvetta seems too much to expect, as dusk is setting in now. While we walk back to the car, the guide plays the song one more time through his speaker. A fraction of a second later, a small brown bird flies across the road just in front of us. We keep an eye on the bush in which it has disappeared.

"There! Top right!"

I quickly put my binoculars to my eyes, and I can see it sitting at the top of the bush. Its distinctive black-and-white head pattern excludes all other species. We burst into cheers.

When we get back to the car it is dark. My Big Year is over. The Black-crowned Fulvetta will go down in the books as number 6,833. My very last species. (Once I return home, it turns out that I forgot to add 19 species, bringing the grand total to 6,852). Exhausted, I plonk into the back seat and put in my earplugs. I close my eyes, press play, and hear Bob Marley's raspy voice. He sings about three little birds who tell him that everything will be fine.

Vlieland, Netherlands (2017)

Far above the Wadden Sea, a group of thrushes flies straight into a violent southwest wind. They fly low to take advantage of the lee troughs. Now and then they skim over the whitecaps. The sky is dark and foreboding, and on the horizon we see a rainstorm that will soon be right over our heads. It's the end of October and I'm doing what I love most: birding, the most beautiful hobby in the world.

Despite the headwind, the group of thrushes is steadily getting closer. I can now see their white eyebrow stripes and ruddy underwing coverts. They are Redwings, originating from the Scandinavian taiga, en route to more southern forests. They have flown over the sea for hours on end now. Their feathers are wet and heavy with rain, and they are at the end of their rope.

One lone Common Blackbird is flying between them. It is an adult male, recognizable by his orange beak and jet-black plumage. Like the Redwings, he is struggling, but his deepest instincts force him to use his last strength.

Finally, they reach the beach, and a little later they land one by one in the only white poplar grove on the northeastern tip of Vlieland. They have a moment to catch their breath, but then they have to move on, because they will be protected from the elements only in the shelter of the forest.

Michiel is standing next to me. "I can't believe how wonderful bird migration is," he says, looking through his binoculars at the exhausted thrushes.

The documentary *Arjan's Big Year* will premiere in a few days. He has worked very hard for the past ten months to make this happen.

Michiel puts an arm around my shoulder. "It's like you never left."

He's right. The year has flown by at breakneck speed, and it still feels like only yesterday that I was standing on my parents' roof terrace and added the Common Blackbird as a sort of number one.

In the meantime, I've had so many incredible experiences. I have crossed the highest mountain ranges, the deepest oceans, the densest rainforests, and the hottest deserts. I have observed nearly two-thirds of all the bird species in the world, from the birds-of-paradise of New Guinea to the albatrosses of the southern oceans to the humming-birds of the Andes Mountains. I have seen nature from its most awe-inspiring side and also from its most fragile side. I saw countless times with my own eyes how we are destroying our planet but also how we are able to learn from our mistakes and turn the tide, even if sometimes all hope seems lost.

When I returned back home to the Netherlands, I received all kinds of inspiring messages: a little boy who had followed my journey closely and gave a lecture on bird conservation, an old man who wanted to include the Preventing Extinctions Program in his will, and countless people who have been affected by my stories and went on to start birding, which brings them an incredible amount of joy every single day. A boy from my high school who used to laugh at my love for birds told me that, after seeing my documentary, he immediately bought himself a bird book and binoculars.

There were, of course, also critical voices. They noted that I flew all over the world and, therefore, left a huge carbon footprint. If I tried to deny this, I would be a hypocrite, because flying is unquestionably detrimental to the environment. But with my trip, I was also able to draw attention to countless conservation initiatives, promote local guides and ecolodges, and raise almost 50,000 euros for BirdLife. This is money that will be used to save the world's most endangered bird species from extinction.

It is important that people continue to visit the natural areas of our planet, because without ecotourism there would not be a tree left in some places. I've seen it with my own eyes over and over again on my trip. Iconic animal species, such as the mountain gorilla and tiger, might have already gone extinct were it not for the fact that they bring in millions of euros every year through ecotourism. Pioneering conservation organizations, such as the Ecuadorian Jocotoco Foundation and the ProAves Foundation in Colombia, rely largely on ecotourism. I would prefer this not to be necessary for the salvation of nature. Why can't we all simply agree we will not touch nature? But unfortunately this is inconceivable at this moment in time, as our modern society has become too far detached from nature.

If you travel, do it in the most responsible way possible and be aware of the negative effects. This applies to all facets of our way of life, from the children we raise to the structures we build, the waste we produce, and the things, food, and drinking water we consume.

***

Above the foredune, suddenly, a Short-eared Owl flies in low. It makes deep strokes with its long, rounded wings and occasionally swings sharply to the left or right to shake off the crows that are chasing it. The evening sun is right behind us, making the owl's light tan plumage stand out, almost golden against the dark rain clouds. It's exactly one year ago that I last saw this beautiful bird, in the snowy winter landscape of Patagonia, on the other side of the world.

As I follow the owl with my binoculars, a black bird with sickle-shaped wings and a deeply forked tail suddenly flutters into my view. My heart rate skyrockets, because a swift in late October is very rare, and the chances to see a Pallid Swift are nearly as slim. I start to shout.

"A *swift*! Above that Short-eared Owl!"

Michiel now also has him in view. Then to our surprise a second bird comes flying next to him. The two swifts fly slowly west over the foredune. One bird makes a somewhat plumper and sandier impression. Total panic breaks out: This has to be a Pallid Swift!

A little later, the identification is complete; this is the twelfth record of this southern European vagrant in the Netherlands. Exactly five years prior, Michiel and I first met when we were standing on a windswept dike near Medemblik, waiting for a Pallid Swift. Back then, my Big Year was merely an idea in my head, and if I'd told him he was going to make a documentary about it someday, he probably would have thought I was crazy.

———

This was the best year of my life. No one can take it away from me. Without the support of my sponsors, friends, family, and above all my parents and Camilla, none of this would have been possible. I am eternally grateful to them. I will also never forget the birders and guides who helped me this year. I can't possibly pick one and say he or she was the best, as almost all of them were awesome. Time and again they went the extra mile. It's thanks to them that this year became such a resounding success.

Have I become a different person after this journey? Absolutely. Most of all, I've become aware of how incredibly privileged I am to have been able to do this. There is so much inequality in this world: The vast majority of people wouldn't be able to make this trip even if they wanted to. Travel, but also nature conservation, is a luxury. It's something that we rich Western people treat rather flippantly. Perhaps this is because we can afford to. However, we have to be keenly aware that billions of people cannot even provide for their basic necessities. Therefore, it is extra admirable that many of them still do everything they can to save nature. Sometimes they do this at the risk of their own life, purely out of love.

To you, reader, I want to say the following: Stay optimistic about the future of the planet, adopt a positive attitude, and do your part. Encourage your children to go outside. Stimulate them by giving them binoculars or a scoop net instead of a smartphone or a computer game, because every child is a nature lover at heart. But above all, enjoy nature and look up to the sky, rather than at your feet or your phone screen. There is so much beauty around you, whether you are in a tropical rainforest or in your own backyard.

RESOURCES

My website
https://arjandwarshuis.com

The documentary *Arjan's Big Year*
https://vimeo.com/ondemand/arjansbigyear

The grand total of my fundraiser (now closed):
www.justgiving.com/fundraising/Biggest-Year

My list following the latest IOC taxonomy
https://world.observation.org/arjan.php

The IOC World Bird List
www.worldbirdnames.org

BirdLife International
www.birdlife.org

Preventing Extinctions Programme
www.birdlife.org/projects/preventing-extinctions

IUCN NL Land Acquisition Fund
www.iucn.nl/en/support-us/support-land-acquisition-fund

Saving the Spoon-billed Sandpiper
www.saving-spoon-billed-sandpiper.com

World Wildlife Fund
www.worldwildlife.org

The Orangutan Project
www.theorangutanproject.eu

Philippine Eagle Foundation
www.philippineeaglefoundation.org

Albatross Task Force
www.birdlife.org/projects/albatross-task-force

Cape Town Pelagics
www.capetownpelagics.com

Alliance for Grassland Renewal
https://grasslandrenewal.org

Rainforest Trust
www.rainforesttrust.org

Ethiopian Wolf Conservation Programme
www.ethiopianwolf.org

Mara Raptor Project
www.mararaptorproject.org

Amazon Conservation
www.amazonconservation.org

Community Cloud Forest Conservation
https://cloudforestconservation.org

Restoring the Peruvian Andes (ECOAN)
www.ecoanperu.org

Jocotoco Conservation Foundation
www.jocotococonservation.org

ProAves Colombia
https://proaves.org

Adopt a Panamanian Rainforest
https://adoptabosque.org/en

Dutch Birding
www.dutchbirding.nl

Observation.org
www.observation.org

ACKNOWLEDGMENTS

First and foremost I would like to thank my parents, Loes and Kees Dwarshuis, and my girlfriend Camilla Dreef. Without your unconditional love, support, and trust, my Big Year and, therefore, this book would never have come about.

I want to express my admiration for Michiel van den Bergh. With *Arjan's Big Year*, you and John Treffer have made a wonderful documentary and a wonderful travel document in a short period of time and with very few resources. In addition, you are a good and critical friend who has repeatedly steered me in the right direction, in the field of planning and communication, but above all in the field of nature conservation.

Vincent van der Spek has also been a huge support. I greatly admire your unparalleled knowledge about birds and your talent for explaining science in an understandable and attractive way to a wide audience. I consider you my mentor. I want to thank Humberto Tan. I don't think there is anyone in the world who has such a passion for other people's passions, a unique and admirable quality to be proud of. You have meant a lot to my Big Year, fundraiser, and documentary, but above all you are just an incredibly nice guy and a great travel companion. Max van Waasdijk, I want to thank you for your company, your perseverance, and of course your friendship.

My Big Year would have been impossible without the right optical equipment, so I would like to thank Marc Plomp from the Bird Information Center of Texel and Gino Merchiers from Swarovski Optik. Without a good camera I could not have documented my trip so beautifully, and for that I would like to thank Wouter Polman and Canon Nederland. I would like to thank Ciska van der Lee and ATPI for their support and professionalism, and for booking all my flights. In addition, I would like to thank Jacqueline de Veer, Caroline Vogel, and ROOTS for your great cooperation and your confidence in my writing abilities. Barend van Gemerden, Martijn Overbeeke, Chris van der Heijden and Vogelbescherming Nederland, Jim Lawrence and BirdLife International, thank you for the fantastic collaboration and hats off for your dedication to protecting our birds. I would also like to thank Rogier Karskens and Bruggink & Van der Velden, Marc Bozon and Coffee3, Fred Dwarshuis and Amsterzonian B&B, Daan Franssen and Café Wester, Sander Bot and Veldshop, Fjällräven, Frank and Birgitta Dreef and Hiensch Engineering, Eddy Zoey, Vrijbuiter, Tim van Oerle and Natuurhuisje, Laurens Steijn of Birdingbreaks, Observation.org, Marten Miske, Kees de Vries, and Dutch Birding. In addition, I would like to thank all my friends, family, and acquaintances who have supported my trip and fundraiser.

This book would never have come about without the right coaching and a fantastic publishing house, so I want to thank Lolies van Grunsven, Denise Larsen, Matthew Derr, Meulenhoff Boekerij, and Chelsea Green Publishing.

Derk Fangman, thank you for your listening ear and your proactive contribution of ideas. A book is not complete without a good title and for that I would like to thank Ton Willemsen.

I would also like to thank all my travel companions, guides, and fellow birders for their friendship, knowledge, flexibility, and dedication. **Netherlands**: Rinse van der Vliet, Garry Bakker, Remco Hofland, Gerjon Gelling, Ies Goedbloed, Jelmer Poelstra, Sander Smit, Arnoud Postema, Joost van Bruggen, Rob Berkelder, Rob Gordijn, Helen Rijkes, Wouter van Pelt, Sjaak Schilperoort, Rob Westerduijn, Loek van Noord, Maurits Holle, Pieter van den Akker, Maurits Munninghoff, Harry Markusse, Tjeerd Burg, André Willem Faber, Jaap Hennevanger, Thomas van der Es, Auke Wibaut, Michel Veldt, Han Zevenhuizen, and Gert Ottens. **UAE**: Mike Barth, Laurent Frangie, and Oscar Campbell. **Sri Lanka**: Jetwing Eco Holidays, Shantha Kumara, and Keith Wijesuriya. **India**: Peter Lobo, Rofikul Islam, Vikram Singh, Sanjay Sharma, Hari Lama, Ikrar Bablu Khan, Deepak Apte, Chewang R Bonpo, and Ranjan Kumar Das. **Thailand**: Petrawut Sitifong and Mark Andrews. **Malaysia**: Andy Boyce and Robert Chong. **Philippines**: Pete Simpson, Zardo Goring, Robert Hutchinson, Bram Demeulemeester, and Rommel Cruz. **Indonesia**: Khaleb Yordan, Malia Tours, and Nurlin Djuni Sorevaya. **Papua New

Guinea: Daniel Vakra, Samuel Kepuknai, Joseph Ando, and Leonard Vaike. **Australia:** Justin Jansen, Gavin Goodyear, Sue and Phill Gregory, Jun Matsui, Els Wakefield, Tim Badwen, Phill, Tim, Phil Peel, Els Wakefield, Tim Nickholds, Peter Waanders, and Ross Jones. **New Zealand:** Johannes Fischer. **Israel:** Itai Shanni, Oz Horine, Meidad Goren, Yoav Perlman, and Nadav Israeli. **Ethiopia:** Merid Gabremichael. **Kenya:** Stratton Hatfield, Susan and Mark Hatfield, Zarek Cockar, Joseph Aengwo, Dixon Lenkoko, Encounter Mara, Nathaniel Mwaumba, and David Ngala. **Uganda:** Ibrahim Senfuma; Raymond, Mathew and Amos Monday Bunengo. **Tanzania:** Furaha Mbilinyi. **Madagascar:** Mamy Randriamanantena, Madagascar Wildlife Tour Agency, and Jacky and Patrice. **Malawi:** Bouke Bijl, Robert, Mvuu Camp, and David Mkandawire. **South Africa:** Ethan Kistler, Billi Krochuk, Callan Cohen, Birding Africa, Errol De Beer, Mathew Axelrod, Jonathan Syke, Zaagkuilsdrift Bird Sanctuary and Lodge, Joe Grosel, Lucky Ngwenya, Jeremy Dickens, Stuart McLean, Klaas-Douwe Dijkstra, and Clifford Dorse. **Ghana:** Mark Williams, Ashanti African Tours, William Apraku, and Paul Mensah. **Spain:** Godfried Scheur, Ecoturex, Lia and Corné van den Bosch, Jose Mazon Hernandez, and Joaquin Mazon Lobo. **Puerto Rico:** Gabriel Lugo. **Suriname:** Sean and Gini Dilrosum, and Fred Pansa. **Brazil:** Eduardo Patrial, Bradley Davis, Bruno Rennó, Bianca Bernardon, Josafa Almeida, and Renato Paiva. **Argentina:** Juan Klavins, Jeroen Martjan Lammertink, Giselle Mangini, Facundo Gandoy, and Diego Monteleone. **Chile:** Sebastián Saiter, Far South Expeditions, Fernando Díaz, and Albatross Birding and Nature Tours. **Peru:** Miguel Lezama Ninancuro, Juvenal Ccahuana Mirano, Wim ten Have, Tanager Tours, and Mr. Moises. **Ecuador:** Dusan and Lorena Brinkhuizen, Olger Licuy, Sani Lodge, Angel Paz, Rolando Sanchez, Wildsumaco Lodge, and Byron Gualavisi. **Colombia:** Jose Castaño, Diego Calderon, Alejandro Pinto, and Birding Colombia. **Cuba:** Angel Martinez Garcia. **Panama:** Guido Berguido, Advantage Tours Panama, Nando Quiroz, and Jan Axel. **Costa Rica:** Pieter Westra, Aratinga Tours, Juan Diego Vargas, Ernesto M. Carman, Christian Urena, Jorge Serano, Eric, Bosque del Tolomuco, Bosque del Rio Tigre Lodge, Abraham Callo, Marcos Mendez, Herman Venegas, and Christian. **Guatemala:** John, Tara, Rob, and Peter Cahill; Lester de Leon; and Max Noack. **Mexico**: Michael Hilchey, BRANT Nature Tours, Larisa Tulasi, Alberto Martinez, RoyalFlycatcher Birding Tours and Nature Photography, Rene Valdez, The Yucatan Jays, Erik Antonio Martinez, Rafael Calderon, Michael Retter, and Mark Stackhouse. **Jamaica**: Wendy Ann Lee. **USA**: Marc Kramer, Eliana Ardila, Carlos Sanchez, Graham Williams, Mike Manetz, Ron Smith, Sparky Stensaas, Gary Nunn, Kaj Dreef, Samantha Bassman, and Philip Chaon. **Japan**: Chris Cook and Marc Brazil. **Taiwan**: Richard Foster. **Vietnam**: Phuc Le.

ABOUT THE AUTHOR

FRISO BOVEN

Arjan Dwarshuis has been fascinated by birds all his life. He works as a professional bird guide and gives lectures and workshops on birding. He starred in the award-winning documentary *Arjan's Big Year* and appears regularly on radio, television, and podcasts in the Netherlands and beyond. He is a columnist for several magazines about nature, and he is committed to the protection of birds around the world. Currently, he is the ambassador of the prestigious IUCN NL Land Acquisition Fund. In 2016, he observed more than 6,852 of the roughly 10,900 bird species in the world, setting a global Big Year record that stands to this day, and that number is still growing thanks to new taxonomic insights. For more information, visit www.arjan dwarshuis.com.

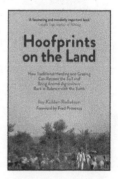
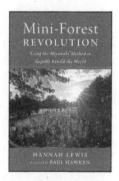

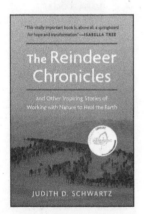